AN ILLUSTRATED
ENCYCLOPAEDIA
OF AUSTRALIAN
WILDLIFE

AN ILLUSTRATED

ENCYCLOPAEDIA OF AUSTRALIAN WILDLIFE

Michael Morcombe

Doubleday & Company, Inc.
Garden City,
New York

Published in Australia by The Macmillan
Company of Australia Pty Ltd, 1974

First published in U.S.A. by Doubleday & Company, Inc.
ISBN 0 385 09596 1
Library of Congress catalog card number 74 15707
© 1974 Michael Morcombe

Printed in Hong Kong

Contents

Introduction

In 1856 when Alfred Russell Wallace travelled from the island of Bali to the island of Lombok, in the archipelago of islands now known as Indonesia, he was impressed by the tremendous difference between the faunas of the two islands. Bali has predominantly the creatures of the Orient; Lombok had a predominantly Australian fauna. Wallace, an English naturalist who was working on a classification of the animals of the world into various natural 'realms', suggested that these narrow Lombok Straits marked the boundary between the Oriental and Australian faunal regions.

Wallace's Line, as it later became known, extends northwards to pass between Borneo and the Celebes, and southeastwards of the Philippines.

Since that time it has been shown that this is not a sharp boundary, but a gradual transition from one region to the other. Many of the mammals of south-east Asia occur on Sumatra, and progressively fewer on each island further eastwards along the chain. Some Asian species extend beyond the line, to Lombok, Timor and New Guinea, while some Australian animals have succeeded in colonizing islands to the west of Wallace's Line. The further an island from Asia, the fewer are its Asian animals — tigers, deer, pigs, monkeys, elephants, rhinoceros, leopards, porcupines — and the closer it is to New Guinea and Australia, the more abundant the creatures of the Australian region, such as possums, cassowaries, cuscuses, spiny anteaters, wallabies, honeyeaters and parrots.

This intermingling of Asian and Australian faunas has occurred only in the last few million years; for a greater period of time Australia and New Guinea were separated from the Asian islands by a far wider expanse of water.

Australia today holds an astonishing array of the strange and the beautiful — the tall eucalypt forests, the profusion of flowering plants, lizards without legs, giant flightless birds, egg-laying and pouched mammals, birds that build incubators and others that build courtship bowers, and a multitude that are plumaged in bright colours.

Australia's fauna is a bizarre mixture of the primitive and the advanced, and it provides a better glimpse of the processes of evolution than any other continental region. The extremely long period of isolation, followed by a time of re-establishment of contact with the rest of the world, has made it possible to observe how groups of animals have been able to radiate into a diversity of forms with minimal outside influence, and how subsequent invasions and superimpositions of other mammals, birds and reptiles have built up into the present rich composite fauna.

Nearly four hundred species of reptiles inhabit the Australian continent, and represent one of the most abundant forms of animal life in the desert regions. One of the most common of the reptile groups is that of skinks, with more than a hundred species which have greatly diversified during their period of isolation.

Geckos, those nocturnal lizards so abundant around the Pacific and Indian Ocean regions, are greatly varied. There are more than forty Australian geckos, some thirty centimetres in length but most considerably smaller.

Other Australian lizards include the dragons, small but in some cases quite dragon-like in appearance. The most startling of these is the Thorny Devil, which has an armour of sharp thick spines over head and body. Another spectacular dragon is the Frilled Lizard, which can suddenly erect an umbrella-like brightly coloured frill around its neck.

Found only in Australia is a group of burrowing lizards which are entirely legless and often very snake-like. These live underground, beneath rocks, forest debris or in dense vegetation where the snake-like wriggling of the body gives better propulsion than small legs. This deterioration of the legs, like flightlessness in birds, has become an advantage within certain habitats; with any major change in the environment such specialization can become an impediment, and lead to extinction.

The most impressive of Australian reptiles are the monitors, or goannas, which although also found outside Australia, have on this continent evolved into their greatest variety. Giants ten metres long which in the past roamed the inland plains have been gone some three million years. But the Perentie remains, an awe-inspiring reptile at close quarters. These big monitors attack their prey with speed and agility, sometimes using their powerful and heavy tails to knock down and stun a victim.

Elsewhere in the world, snakes are predominantly of the colubrid group, either non-venomous or only moderately venomous. However in Australia the very deadly elapid snakes are the most abundant. The Taipan, often more than three metres in length, and very fast-moving, is more deadly than the King Cobra, and one of the most dangerous snakes in the world.

Australia has a rich and varied bird population. Here are found magnificent Black Swans with crimson beaks, muttonbirds in flocks of thousands, Brolgas in stately courtship dances, parrots and cockatoos dashing through the treetops displaying brilliant colours, bowerbirds attending courtship arenas and the megapodes their egg-brooding mounds. Here are honeyeaters in great numbers probing the flowers, and huge flightless birds in the jungles and on inland plains.

But it is probably for its mammals that the Australian region has become best known — the egg-laying monotremes and the pouched marsupials are almost symbolic of Australia throughout the world. Besides those favourites, the kangaroo, koala and platypus, there are many, many others, equally bizarre in appearance, as strange in habits, and far less well known — the tiny but fierce marsupial-mice which hunt spiders and scorpions at night, the potoroos, pademelons, bettongs and Quokka, possums and gliders, kangaroo-mice, a host of bats large and small, and even marsupials that spend their lives beneath the ground.

Reptiles

The Australian reptile population is a composite of ancient types, and comparative newcomers which are more closely allied to the present-day reptiles of other continents. Some have been on Australia since its origin as a separate continent; others have come from the north, via New Guinea. Any falls of sea level greater than about ten metres below the present level have connected Cape York to New Guinea. A deeper channel between the islands of Bali and Lombok has not been as effective a barrier to reptile crossings as it was for mammals. Some of the reptilian immigrants came when these lowered sea levels formed land bridges, and others by 'island hopping'. The great number of islands large and small, and the tiny coral cays and exposed reefs, were colonized each in turn by reptiles that had swum, drifted or been blown across intervening channels, until the Australian mainland was reached.

Today many Australian reptiles are close relatives of those found in Asia or elsewhere around the world. Here they live side by side with lizards and snakes that have been on this island continent so long that their adaptations to the Australian environment have made them quite unique.

Within Australia, a rather limited utilization by the mammals of the tremendous variety of environmental 'niches' has encouraged the evolution of a greater diversity of reptiles. The extreme aridity of central regions of Australia has long been a barrier to the southwards dispersal of new arrivals from the north; for these species, the forests down the eastern seaboard have been a corridor to southern Australia.

The more 'recently' arrived species, such as the forest dragons, are still tied to the north-eastern rainforests which, of all Australian habitats, most closely resemble their New Guinea home. It is probably within Australia's desert reptilian fauna that are found the most distinctive species, well adapted for survival in that harsh environment.

By world standards, Australia, for its size, supports more than its quota of reptiles, although some have failed to reach or become established here.

A few groups have no living relatives elsewhere in the world. Among these are the legless pygopodid lizards, which resemble snakes, but still retain vestigial remnants of hind legs looking like larger scales projecting from the sides of the body.

Other Australian reptiles, tend to have their closest family ties either in South America (especially the tortoises), or in Asia (geckoes, skinks, crocodiles, monitors and dragons).

Among the oldest and most impressive lizards of Australia are the big predatory monitors, commonly known as goannas. Fifteen of the world's twenty-four species occur in Australasia, including the second-largest, the Perentie, which attains a length of almost three metres.

Reptiles have certain characteristics in common which distinguish them from other forms of animal life. Obvious external features are the usually scaly skin, and the squatting appearance, with legs protruding almost horizontally from the sides of the body, which lies on or very close to the ground.

Internally, reptiles differ from birds and mammals in several respects, and this has a great influence upon their behavior and capabilities.

Instead of generating and maintaining a constant temperature by internal means as do birds and mammals, reptiles must rely greatly upon warmth from the environment. They must spend a great deal of time following patterns of behaviour which will regulate their body temperature, basking in the sun in the morning until warm, then retreating to shade, burrow or crevice when too hot. Although their body temperatures fluctuate, they cannot strictly be called 'cold-blooded' as their preferred temperature is near our own.

Reptiles, other than crocodiles, have a relatively inefficient circulatory system, the ventricle being incompletely divided and forming in effect, three, instead of four heart chambers as found in birds and mammals. This allows some intermixing of arterial and venous blood, resulting in a lower level of oxygenation of the blood reaching the muscles. Although at times capable of very rapid action, reptiles cannot sustain such peaks of activity for very long periods.

The sense of smell is extremely well-developed in many lizards. The monitors, for example, always forage for food and investigate new surroundings with their tongues flickering out.

To smell substances picked up by the tongues there are, in addition to the nose, special sense organs situated in the root of the mouth towards the front of the snout. Known as the organs of Jacobson, these hollow ducts receive the tips of the tongue, and detect faint odours collected on the moist tongue tip.

Many reptiles have very keen eyesight. Exceptions are those that spend most of their time beneath the ground. The structure of reptiles' eyes is influenced by their way of life. Those which are nocturnal generally have elliptical pupils, while in diurnal species the pupil is circular.

Some lizards have remarkably well-adapted eyes which can function efficiently both at night and in full sunlight. Although extremely sensitive, these eyes are not blinded by very bright light. The aperture of the pupil can be closed down until the opening virtually disappears. With some species the slit-like aperture may become a line of pinhole openings, giving definite optical advantages, such as a deeper zone of sharp vision (in photographic terms, a greater 'depth of field').

Reptiles, but for a few (like the Perentie) large enough to fear no natural predator, have evolved various means of protection. Bluff is very important in deterring a potential enemy, but other reptiles rely more upon camouflage, mimicry, speed of escape, or other sometimes bizarre but effective survival tactics.

Defensive display behavior is common among Australian lizards. The monitors, which rear upon hind legs, then appear much more fearsome. Some species have developed special appendages which can suddenly be erected, giving a startling increase in size. The Frilled Lizard provides the best example of this, the coloured frill exploding outwards at the same instant that the reptile gapes its bright yellow mouth at its enemy.

Many of the smaller lizards mimic dangerous species. The flap-footed ('legless') lizards imitate in their colouration, shape and behaviour, small venomous snakes.

Some lizards, especially the skinks and geckoes, are able, voluntarily, to break off portions of their tails, so that if caught by a predator (and the tail is the most likely point of capture of a fast-fleeing lizard) they are able to break free, and escape while the twitching tail serves as a distraction.

The lizard is actually able to control where the break occurs. Sudden contraction of muscles causes the tail to break at one of many predetermined planes of weakness in the tail vertebrae; generally the break will be made where the tail loss is the minimum needed to escape. A re-grown tail differs in appearance from the original, and cannot be shed within the new-grown portion.

BLIND SNAKES

FAMILY *TYPHLOPIDAE*

These worm-like and very small snakes spend almost their entire lives underground, so that their eyes are now no more than dark spots under the semi-transparent scales. Their bodies are smooth and uniform in thickness, and the scales polished, allowing them to slip through the soil with a minimum of resistance. At the tip of the blunt tail is a small spine which is pushed against the soil to help the blind snake push its way forward.

At times, almost invariably at night or after heavy rain, blind snakes may be seen on the surface of the ground. Some species are found most commonly in termite mounds, where they have an assured supply of food.

Blind Snake   *Typhlops*

Also known as worm snakes, these very small and completely harmless snakes have a superficial resemblance to an earthworm, but with a more glossy appearance, being covered with tiny shiny scales.

After many thousands of years of subteranean living their eyes have degenerated until now they are no more than dark spots beneath the transparent ocular scales. It is probable that they can do no more than distinguish light from dark, sufficient to ensure that the snake avoids exposing itself in daylight when it would be completely vulnerable.

Blind Snakes are burrowers, the tip of their snout being hard and pointed, and the mouth recessed on the undersurface so that they can push through the soil and decayed logs where they prey upon termites and ant eggs.

THE COLUBRID SNAKES

FAMILY   *COLUBRIDAE*

Although extremely widespread and abundant in other parts of the world, the colubrid snake family has few representatives within Australia. The group contains both venomous and non-venomous species, those that are venomous being known as 'rear-fanged' snakes, their fangs being towards the back of their mouths. Compared with the elapid snakes, their venom-injecting apparatus is much less efficient, and generally they are not considered to be dangerous. Those members of the colubrid family which are non-venomous (such as the Green Tree Snake) may simply swallow their prey whole and alive, or constrict by a tightening action of the body coils in a manner similar to the pythons.

Green Tree Snake
Dendrelaphis punctualatus

Although non-venomous, the Green Tree-snake is aggressive, and will attack and bite to defend itself. In color it varies from vivid green to dark olive-green above and from yellow to almost black beneath; a consistent feature is the sprinkling of bright blue spots which appear as the

snake, when angered, expands its body and reveals blue between the scales.

This long slender snake, as its name suggests, is almost invariably seen in trees, where it can move through the foliage with a fast, graceful flowing motion. The Green Tree Snake has a wide distribution through the forests of tropical northern and north-eastern Australia.

CROCODILES

FAMILY   *CROCODYLIDAE*

Crocodiles are characterized by massive, muscular and heavily armoured bodies, and by their many modifications for the aquatic life. The armour consists of bony plates just below the surface of the thick leathery skin of the back, while the abdominal region is partly protected by a series of rib-like bones, which are not attached to the rest of the skeleton. The very heavily built skull is well equipped with pointed teeth that are used for holding and tearing of the prey. The large and powerful tail, which is flattened laterally, drives the crocodile through the water, all four limbs trailing against the sides of the body.

The prey, particularly a larger animal, is killed by drowning, the crocodile being able to hold the prey beneath the surface yet still continuing to breathe itself. The nostrils are raised and situated at the very tip of the snout, and are equipped with valves which shut when the crocodile submerges. There are other modifications to the air passages from nose to throat to enable the crocodile to breathe while its mouth is open (to hold its victim) but nearly submerged.

Crocodiles lay eggs, which in the case of the Saltwater Crocodile are placed in a nest of grass, leaf-litter and mud on a river bank above high water level. The nest is covered with vegetation (which in decaying generates heat to assist incubation) and mud. The female crocodile remains in the vicinity to repair and protect the nest, and possibly to assist the newly hatched young to escape from beneath the mud.

Freshwater Crocodile
Crocodylus johnstoni

Australia has two species of crocodile. The sea-going Eusurine or Salt-water Crocodile, is widely distributed northwards from Australia to New Guinea, Malaysia, India and southern China, while the smaller Fresh-water or Johnstone's Crocodile is found only in the rivers of northern Australia.

The Freshwater Crocodile grows to a maximum length of about three metres. Since it has been made a protected species in tropical Western Australia and the Northern Territory this crocodile may have become more common, for it can readily be seen in the river pools of some of the northern national parks.

The relatively small Freshwater Crocodile is completely harmless to man, feeding upon fish and other aquatic life. Crocodiles catch such prey as water birds by approaching stealthily underwater, and seizing from below. At other times they wait submerged (just their raised nostrils and eyes showing), looking in shape and texture just like an old waterlogged log.

Although spending most of their time in the water, crocodiles often climb onto sand banks or rocks to bask in the sun. Here they remain very alert and splash into the water at the first sign of anything unusual.

Saltwater Crocodile   *Crocodylus porosus*

The only large indigenous animal that is potentially dangerous to man, the Saltwater or Esturine Crocodile grows to an enormous size: specimens of more than ten metres in length having been recorded. However the past extensive slaughter of crocodiles in northern Australia, principally for their skins, makes it most unlikely that many specimens of that size, which could be up to two hundred years in age, survive today.

Crocodiles, unlike the relatively docile alligators, are extremely savage, and even specimens

---

Green Tree Snake   *Dendrelaphis punctulatus*
Capable of moving through the foliage of trees and undergrowth with incredible speed, the long, slender and extremely agile Green Tree Snake has specially ridged belly scales that assist it to glide around tree-trunks and along branches. It spends most of its time off the ground, and although not venomous will strike aggressively if disturbed.

Blind Snake   *Genus ramphotyphlops*
Living in the permanent darkness beneath logs, under stones or in the soil, the small, harmless blind snakes, of which there are many species, have only dark spots for eyes. When disturbed they may tie themselves into tight knot-like bundles and emit a strong and very unpleasant odour.

Saltwater Crocodile   *Crocodylus porosus*
A small Saltwater Crocodile demonstrates the viciousness of the species — a ten metres long specimen would best be treated with respect! Crocodiles can be differentiated from alligators, which occur only in the Americas and China, most easily by observing the long fourth tooth of the lower jaw. In the case of the alligator this tooth fits into a socket in the upper jaw, and is not visible when the mouth is closed; with crocodiles it fits into a notch on the outer side of the upper jaw, and therefore is always visible.

Bearded Dragon   *Amphibolurus barbatus*
The Bearded Dragon when alarmed or annoyed opens its mouth threateningly and puffs out its 'beard' of spiny scales. At the same time it inflates its body with air, and endeavours to look as big and as dangerous as possible.

that have been many years in captivity retain their vicious temperaments. There are many authentic records of man-eating attempts both successful and unsuccessful, by this species.

The Saltwater Crocodile is generally found in coastal estuaries and the lower reaches of rivers. It commonly travels in the open sea, and is widely distributed, from Australia northwards to New Guinea, Malaysia, India and southern China.

THE DRAGONS

FAMILY *AGAMIDAE*

The lizards of this family are widely distributed from southern Europe, Africa and Asia, to Australia where a great variety of species occur. Certain features separate the dragons from other lizards. The teeth have become well differentiated according to their functions, especially in the larger species, which have enlarged canine and incisor type teeth at the front of the mouth; the teeth of most other lizards are rather uniform in shape and structure. The dragons have tails which although sometimes long and easily broken, are not voluntarily shed; regrowth of the tail is infrequent, and then only partial. Their scales lie side by side on the body, not overlapping as with some other lizards, and are rough, dull and often spiny.

Bearded Dragon   *Amphibolurus barbatus*

The fearsome looking Bearded Dragon is one of those reptiles able to change its colours. While individuals from coastal areas are generally greyish, those of the deserts of the interior are yellowish-brown or red-brown, but able to become brighter or darker, turning almost black on the head, when the lizard is excited, or trying to escape detection by matching its surroundings. Possibly the colour changes can also be used to increase the rate of heat absorption when basking in the sun after a cold night.

Being a large reptile, up to about sixty centimetres in length and stoutly built, the Bearded Dragon takes longer to warm up in the morning. Before it can become really active it must bask in the sun for a long period, and to expose the maximum surface to the sun it finds a warm place, such as a rock or road surface, and flattens out its body to a remarkable degree to catch more of the sun.

The Bearded Dragon rarely tries to escape by running, but relies upon its cryptic coloration and the spiny broken body outline for camouflage. Lying perfectly still, it is very hard to see and generally overlooked in its natural environment.

Boyd's Forest Dragon   *Goniocephalus boydii*
Although awesome in appearance, Boyd's Forest Dragon is a docile and completely harmless creature which depends upon concealing colours and spiny armour for its survival in the rainforests of north-eastern Queensland.

Boyd's Forest Dragon
Goniocephalus boydii

In the gloomy mossy jungles of north-eastern Queensland the Forest Dragon's greenish colours and motionless posture on limb or tree-trunk make it very difficult to see. Despite its awesome appearance it is a docile and harmless creature, an insect-eater, which depends upon its concealing colours and spiny armour-plating for survival.

The Forest Dragon is a representative of one of the most recent groups of reptiles to enter Australia from New Guinea. The three Australian species (the other two being *Goniocephalus godeffroyii* of Cape York, and *G. spinipes* which extend southwards to New South Wales) are still tied to heavy forest country which most closely resembles the New Guinea environment.

Boyd's Forest Dragon is said to be a rare species, but this may result from its superb camouflage. It grows to a length of about fifty centimetres of which about half is tail.

Eastern Water Dragon
Physignathus lesueurii

The water dragons are the only dragon lizards in Australia that are completely dependent upon an aquatic habitat. The Eastern Water Dragon is found only along the creeks and rivers of eastern Australia from Cape York to southern New South Wales. These large reptiles, coloured dark brown with black blotches along the back and a red undersurface, can often be glimpsed lying on logs or on the slopes of river banks. But immediately one catches a glimpse of any intruder it drops into the water, and may remain submerged for as long as half an hour. These reptiles show special adaptations for life in the water.

The tail is flattened laterally, and with waving undulations provides propulsive power through the water, while the legs remain folded back along the body.

A distinctively coloured subspecies, the Gippsland Water Dragon (*Physignathus lesueurii howittii*) occurs in eastern Victoria. It has bright blue and yellow throat markings, and less spiny scales.

Frilled Lizard   *Chlamydosaurus kingii*

Probably the most spectacular and colourful of all Australian lizards, the Frilled has developed to the full the survival tactics of surprise and bluff. The abruptness of the display is all important. At one instant the lizard is running — and it is very fast, dashing along upon its hind legs — then suddenly it has turned on its foe, and explodes into a shape that, from the front, is now at least four times as big, with warning colours of red and orange, and a gaping yellow mouth. For the predator that was about to seize a slender lizard that was fleeing for its life, the transformation cannot fail to be alarming, if but for a few seconds. But in that moment of reprieve the fleet-footed Frilled Lizard is likely to have attained the safety of a tree or rock crevice.

The Frilled Lizard has a wide distribution across northern Australia, and the colours of the frill membrane vary considerably, being usually yellow with black and white markings in central Queensland, and orange with reddish, black and white markings in the Northern Territory and Kimberleys.

Ornate Dragon   *Amphibolurus ornatus*

This quite large (25 centimetre) dragon lizard inhabits rocky outcrops, where it lives beneath flat exfoliated slabs. In these situations in south-western Australia it is very common, and attracts attention to itself by running at great speed across the granite surfaces, stopping now and then to watch, then dashing away again, or diving into a crevice. The coloration is variable, but consists mainly of black and yellow markings.

More than a third of Australia's dragon lizards belong to the genus *Amphibolurus*. Unlike skinks, which generally have glossy scales, dragons have rough scales; the body is usually flattened or laterally expanded, and the tail is long and not readily shed.

Streaked Earless Dragon
Tympanocryptus lineata

The small lizards of this genus are distinguished from all other members of the dragon lizard family by the complete absence of any external ear opening. The generic name, *Tympanocryptus*, refers to the hidden tympannum, or ear drum membrane. On other reptiles this is visible as a more or less circular or ellipticial disc on the sides of the head or recessed into an ear opening.

The streaks consist of five fine white lines along the body. There are two colour forms of this dragon, the most typical having a light brown head, and brownish body with about six darker cross-bands, and additional bands across the tail; the other form has a paler head with rufous patches across the snout, in front of the eye and across the nape of the neck, while the back is greyish with wide brown crossbands.

The Streaked Earless Dragon, which grows to no more than fourteen centimetres in length, inhabits the dry interior of Australia.

Thorny or Mountain Devil
Moloch horridus

Probably the most bizarre of Australia's reptiles is the ant-eating Thorny Devil. Apart from its more obvious peculiarities — the hard spines that cover head, body and limbs, the peculiar spiny hump above its neck, its apparent confidence in camouflage, and fortress construction in miniature as a means of defence, the Devil has many other remarkable characteristics.

Most amazing, and most useful for a desert dweller, is its ability to drink through its skin. Dews are common when the temperature of the clear desert air drops sharply after sunset. The film of moisture of dew, or the brief wetness of a thunderstorm can be utilized by the Mountain Devil. The water is not absorbed into the skin tissues, but passes into tiny capillary channels leading to the mouth, where it collects and is swallowed.

The Mountain Devil has the ability, within limits, to change colour, by varying the concentration of yellow, red and black pigments, to match the colour of the desert sand or clay surface.

Western Jew Lizard
Amphibolurus barbatus minor

A subspecies of the Bearded Dragon, this dragon is distinguished from the eastern typical lizard by the noticeably smaller 'beard' of spiny scales, and by its smaller size, and more slender build. This form inhabits the western half of the continent; its colour is quite variable, ranging from an almost uniform dark brown to a much lighter tawny brown.

Western Netted Dragon
Amphibolurus reticulatus

This colourful dragon lizard, which grows to a length of about twenty-five centimetres, occurs in the deserts and dry scrublands of the western part of the continent; eastwards, it is replaced by the Central Netted Dragon. The coloration of this species shows considerable variation as well as sexual differences. The adult male has darker and lighter head markings which are often obscured by the dust of the soil in which it burrows. The back and sides are covered with a dark network of lines enclosing pale yellow spots, the reticulated pattern from which the lizard derives its name.

The female has as its main pattern a number of dark patches arranged symmetrically down its back; the reticulation markings are only faintly visible along the sides. Colour markings of the young are so varied that they cannot be used as a reliable means of identification. The Western Netted Dragon lives in shallow burrows usually in sand dune country.

THE ELAPID SNAKES

FAMILY *ELAPIDAE*

More than half of the land snakes in Australia belong to this family which contains the world's most venomous snakes; the venom of the Australian Taipan and Tiger Snake is more potent than the cobras of Africa and Asia or the mambas of Africa. The elapid snakes possess a pair of fangs at the front of the top jaw, each fang having a hollow or deep groove through which the venom is ejected.

The venom of these snakes contains a complex mixture of proteins, including haemorrhagins, which destroy blood vessel linings allowing blood to escape, neurotoxins which act on the nervous system to cause paralysis, haemolysins which destroy red blood corpuscles, thrombase which causes blood clotting, and other substances. The proportions of each vary from one species to another. The elapid snakes have powerful neurotoxins, which take rapid effect.

Fortunately the fangs of most species are short (so that thick clothing gives a measure of protection) and few are aggressive unless cornered or attacked. The majority of species are small and feed upon insects, lizards and other small prey; their bite can be painful but not dangerous. Some elapid snakes lay eggs, while others produce live young.

Bandy Bandy   *Simoselaps bertholdi*

A small colourful burrowing snake, generally no more than forty centimetres in length. It is harmless; its fangs are so small that they would probably not penetrate human skin. Its prey consists of insects, frogs and small lizards.

The Bandy Bandy, which was greatly feared by the aboriginals, is strikingly patterned, having a number of almost black cross bands which are as wide as the coloured interspaces. Most of the bands continue as complete rings around the body. The colour is variable, the head being greyish, while the body between black rings may be reddish, orange or creamy yellow. The undersurface is pale yellow, or almost white.

Black Striped Snake   *Vermicella calonota*

This small snake is one of the most colourful Australian reptiles. Although venomous, it is not harmful to man, and lives upon termites and similar small creatures. It is a burrower, and has a sharp-pointed snout (as seen from the side) which projects over and protects the mouth. The range of this species, according to specimens at the Western Australian Museum, appears to be restricted to the country within about thirty-five kilometres of Perth.

Copperhead   *Denisonia superba*

Although the colours of this venomous Australian snake are variable, the pale yellowish edgings which give a striped appearance to the lips are a characteristic feature.

Copperheads are found in swampy grassy areas of mountainous south-eastern Australia and Tasmania. In the northern parts of their range of distribution they are often reddish on the upper surface and yellow below; in the Blue Mountains they are reddish brown or almost black, while in southern Victoria the colours vary from a light coppery brown to dark brown. Large individuals of this species may be up to two metres in length, and although deadly, fortunately prefer to retreat rather than attack.

Death Adder   *Acanthophis antarcticus*

This is one of Austrlaia's most highly venomous snakes, yet rarely attains a length of as much as one metre. It is broad and flat, with a wide triangular head that is very distinctly separated from the body by a narrow neck. The short, thick body terminates in a peculiar thin little tail, which has a whitish spine-like tip. The colour of the death adder is variable, but generally greyish to reddish-brown with numerous darker crossbars.

The peculiar tail appears to function as a lure, the Death Adder lying in a coiled, ready-to-strike position with the tail tip in direct line with its head. If a lizard or other small creature approaches the tail tip is wriggled and the lizard that ventures close to capture what appears to be some insect or other live food becomes, in one lightning fast strike, the prey of the Death Adder.

Although this snake is not aggressive, it is likely to hit at anything coming within range of where it lies, and then it strikes with great speed and accuracy. As it can be extremely well camouflaged, there is always some danger of stepping on or close to one. Although the death rate from its bite is high, there is a good chance of recovery if there is prompt first aid and subsequent medical attention.

The Death Adder has a wide distribution in all States except Tasmania; a very similar species, the Desert Death Adder, *A. pyrrhus*, occurs in central and north-western Australia.

Little Whip Snake   *Denisonia gouldii*

This colourful and only slightly venomous little snake, which grows to a maximum length of about sixty centimetres, is usually found in roots, firewood, ant nests or termite mounds, where it feeds upon various small insects.

Although typically coloured reddish-orange with a sharply contrasting black head, and creamy to pinkish-white beneath, there is also a dark form on which the entire back is very dark brown or black. There are two creamy-white marks just in front of the eyes. The Little Whip Snake is confined to the south-western corner of the continent.

Although this small snake is venomous, it is a beneficial species, and certainly not dangerous.

Mulga Snake   *Pseudechis australis*

A large brown snake, growing to a length of almost three metres, and averaging two metres. It is distributed across Australia from Queensland through the Northern Territory to Western Australia, coming as far south as the northernmost parts of South Australia and Victoria. It inhabits mostly the dry mallee eucalypt and mulga (acacia) scrub country.

Because of its large size and brownish colour the Mulga Snake, also known as the King Brown Snake, is sometimes confused with the deadly Taipan. It is related to the black snakes, and because it is the second largest of Australian venomous snakes, the bite of a large specimen could possibly be fatal. Unlike the majority of snakes, the Mulga hangs on when biting, injecting a great quantity of venom.

Taipan   *Oxyuranus scutellatus*

One of the largest of Australia's venomous snakes, and certainly the most deadly. Before the introduction of an antivenene recovery from its bite was unlikely. Although its venom may be very slightly less toxic than that of the Tiger Snake, which is one of the most potent in the world, the Taipan's centimetre long fangs inject the venom much more deeply, and in much greater quantity.

Although the Taipan in its natural haunts is shy and elusive, rarely attacking unless aroused, it becomes extremely ferocious when threatened. Being a long slender snake it can strike from a considerable distance, and often drives the fangs into several places in rapid succession.

Taipans occur in coastal north-eastern Australia, from the Gulf of Carpentaria and Cape York southwards to Gympie, towards south-eastern Queensland.

Taipans generally measure between two and three metres in length; their colour is coppery brown to dark brown above, and yellow speckled with orange on the undersurface. In the far north it could be confused with the King Brown Snake, *Pseudechis australis*, which also attains a length of up to three metres.

Tiger Snake   *Notechis scutatus*

This is considered to be the most dangerous of Australian snakes in the southern states, although it almost always prefers to retreat rather than attack. Principally an inhabitant of swampy areas, it feeds largely upon frogs; it is partly nocturnal in its activities.

The colours of the Tiger Snake vary greatly, and may be yellowish, brown or black, generally with dark transverse bands. The Western Tiger Snake, of south-western Australia, is steely blue-black, sometimes with many narrow orange or yellow cross bands; the undersurface is pale yellowish. When disturbed the Tiger Snake flattens its body to a remarkable degree, the neck region in particular taking on a cobra-like hooded appearance.

The venom of the Tiger Snake is one of the most potent types known, but fortunately the quantity injected at a bite is small. However any large Tiger Snake must be considered to be potentially lethal, and best avoided rather than provoked into attack.

Yellow-faced Whip Snake
Demansia psammophis

This slender snake, which inhabits the coastal sandstone country and sandy inland areas of south-eastern Australia, grows to a length of a little more than one metre. It is greyish above and near white below, while the eye is outlined in yellow.

Although it is one of the elapid snakes (a family which contains Australia's most deadly snakes) the Yellow-faced Whip Snake is not considered dangerous. A bite would cause discomfort, but not death to a healthy adult; the bite of an average-sized Whip Snake would probably be no worse in its effect than a hornet sting.

THE GECKOES

FAMILY   *GEKKONIDAE*

Australia has about fifty species of geckoes, most of which are small, less than ten centimetres in length. They are almost exclusively nocturnal, and this is reflected in their possession of very large eyes, and delicate flabby skin on which the scales are so small and widely separated that the appearance is of a finely granular surface. Some species have the belly skin so transparent that the internal organs are visible. The scales are often modified to form spines or tubercules. Gecko eyes have a vertical slit pupil, which at night dilates almost to the full size of the eye, giving excellent vision in very dim light. The eyes have no lids, but are covered by a transparent scale, and are cleaned with a lick from the long flat tongue. Most geckoes are ground-dwellers, but some have their toes flattened and with adhesive pads which enable them to walk on smooth vertical surfaces, including glass. The toe pads are not sticky, but have thousands of fine hair-like bristles which are probably pressed into microscopic irregularities in the surface to give a grip when climbing.

Barking Gecko   *Phyllurus milii*

A widely distributed small gecko which grows to a length of about fifteen centimetres, the Barking Gecko has a granular-surfaced skin, across which are scattered small conical tubercules. These are white and in bold contrast against the dark body colour, giving the gecko a white-spotted appearance.

The Barking Gecko possesses the large eyes typical of the nocturnal geckoes. It is one of the commonest of geckoes in southern Australia, and takes its name from the little cough-like barking noise it makes when alarmed.

Broad-tailed Rock Gecko
Phyllurus platurus

Also known as the Southern Leaf-tail Gecko, this species inhabits sandstone rock exfoliations in northern New South Wales and southern Queensland. It is common in the Sydney district.

The colour of ground-dwelling geckoes usually resembles that of the soil of the area, while patterns are arranged so that they are perfectly camouflaged in their natural environment. Tree and rock-dwelling species are usually brighter in colour than ground dwellers. Geckoes eat a great variety of food, and their prey must be moving because they seem unable to distinguish motionless objects. Insects and spiders are the principal food, but scorpions, centipedes and occasionally smaller lizards may be taken. The food is crushed then swallowed whole. Water is obtained from moist surfaces such as leaves or bark.

Leaf-tailed Gecko   *Phyllurus cornutus*

The Leaf-tail provides one of the most perfect examples of camouflage in the reptile world. Its colours blend perfectly with the mossy bark of the trees of its rainforest habitat, so that it is almost invisible during the day when lying pressed flat against tree-trunk or branch. Even the contours of its body outline are broken by spiny shapes, while its flattened edges ensure that it casts a minimal shadow outline. The hollows of jungle trees, such as the huge figs and giant stinging trees, and the rocky crevices around the entrance of caves, are favoured homes for the Leaf-tail.

At night this gecko comes to life, and preys upon insects, spiders and smaller lizards, which

are captured by a slow, cautious stalking approach and a sudden final forward lunge.

Smooth Knob-tailed Gecko
Nephrurus levis

A burrow-dwelling gecko with huge eyes, this gecko has a large tail which is used for fat storage as a reserve of food for drought seasons, for this too is an inhabitant of the dry interior of Australia.

The Smooth Knob-tailed Gecko is very attractively coloured, being reddish brown with white or creamy spots on body and tail, and delicate lilac tints on the head.

Unlike the Spiny Knob-tailed Gecko, this species is able to shed its tail. This ability, known as autotomy, is very well developed among geckoes, which readily break off their tails if not handled gently. Autotomy is under the voluntary control of the lizard, which can control the position of the break. A new tail grows later, and generally differs in scale pattern and colour. Because the new tail is built around a central rod of cartilage it cannot be shed within the re-grown portion. All subsequent breaks must be of a larger piece of tail, broken through the vertebrae which are designed to snap under the strain of certain strong muscular contractions.

Spiny Knob-tailed Gecko   *Nephrurus asper*

One of the most bizarre small lizards in Australia, with its big head, stout body and ridiculously short tail, this gecko differs from most others in another respect — it is unable to shed its tail. This presumably results from there being almost no tail for a predator to grip. Although with a length of no more than about 1.7 centimetres the Spiny Knob-tailed Gecko has the greatest bulk of all Australian geckoes, as so very little of this length is tail. It is brownish in colour.

This species from inland Australia has been observed to have a remarkable pattern of display behaviour, the limbs being extended and then retracted continuously for several minutes.

When confronted by a small predator such as another lizard of similar or only slightly larger size, this knob-tailed gecko threatens by widely opening its mouth, which is huge for the size of the animal, and lunging forward in mock attack. But if the predator is very much bigger — a large snake or goanna — the gecko wisely wastes no time on threats which experience must have shown its ancestors to be useless on such occasions. Without hesitation it turns and runs.

THE GOANNAS OR MONITORS

FAMILY   *VARANIDAE*

Included in this family are the world's largest lizards; Australia has about twenty of the world's total of twenty-five species of goannas. At least fifteen are restricted to Australia, where the largest species is the Perentie, over two metres in length. The goannas are all powerfully built, well equipped for a predatory and scavenging role with long sharp claws, and teeth that are sharp, long, backwardly curved and very numerous. Their prey consists of snakes, lizards, small mammals (including rabbits), birds and insects; they will also feed on any carrion found. Most are ground-dwelling, but some species climb trees when hunting or to escape an enemy.

Lace Monitor   *Varanus varius*

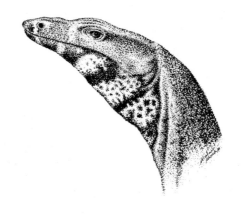

This is one of the largest of the Australian monitors, its two-metre length exceeded in bulk only by the Perentie. The heavily built body and powerful limbs are immediately obvious. Like the other large monitors, it is carnivorous, but will also eat any carrion it may find.

There are two colour phases of the Lace Monitor. Most commonly it is black with yellow patches or indistinct yellow bands; the other form is more colourful, with bright bands of yellow or orange which may be wider than the black. Young specimens are generally more brightly coloured.

The Lace Monitor has a wide distribution in eastern Australia; it does not inhabit the arid interior or the far north. A hollow log, a hole in the ground, or leaf litter may be the site for concealing the eggs, up to twelve in number.

Perentie   *Varanus giganteus*

This massive goanna, which attains a length in excess of two metres and may weigh as much as ten kilograms, is a predator which will eat anything it can capture. Its prey consists of snakes, lizards, ground-dwelling birds, and mammals, even small kangaroos. Rabbits, caught in their burrows, now comprise a major part of its diet, making this large monitor a most beneficial species.

The Perentie inhabits desert and semi-arid regions, and occurs in all mainland States of Australia, except Victoria. It lives in burrows or beneath rocks.

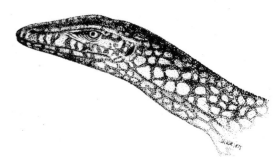

Generally the Perentie, despite its formidable size, will endeavour to avoid danger by running, heading for a tree where, well above the ground, it adopts a rigid pose, and with its colour pattern disrupting the body outline, can easily be mistaken for a branch. The big Australian monitors never attack humans, but if cornered and threatened will defend themselves. For this they are well equipped, having massive claws and powerful jaws, while the lashing of the heavy tail also can cause considerable damage.

Desert Pygmy Monitor   *Varanus eremias*

Very widely distributed through central parts of South and West Australia, the Desert Pygmy Monitor is a small ground-dwelling goanna which favours the desert sand-dune country, where it lives among the clumps of needle-leaved porcupine-bush or spinifex. In this warm climate it is active throughout the year, and travels as much as over half a kilometre in a day in search of its prey, principally large insects such as grasshoppers, but also various small lizards.

The Desert Pygmy Monitor lays three or four soft parchment-shelled eggs. Many of the Australian *Varanus* monitors choose termite mounds as natural incubators — the female scratches into the mound and deposits her eggs.

Frilled Lizard   *Chlamydosaurus kingii*
Clinging to the leafless top of a dead pandanus palm, a Frilled Lizard endeavours to look bigger and more dangerous by expanding the frill of bright warning colours, which normally lies folded around his neck. The wider he gapes the more the frill stands out, for it is supported on long slender extensions of the hyoid bones which move as the mouth is opened. If his bluff works, and his enemy hesitates a moment, the lizard will drop to the ground and run at great speed to a higher and more secure tree-top retreat.

Thorny Devil *Moloch horridus*
One of Australia's strangest reptiles, its body covered
with long spines, and a large hump (the function of
which is unknown) on the back of its neck, the Thorny
or Mountain Devil is an inhabitant of the inland and
desert parts of Australia. Unlike most other members of
the dragon family, this creature is very slow-moving,
and spends much time beside ant trails, flicking the
ants one at a time into its mouth.

Streaked Earless Dragon *Tympanocryptus lineata*
This small colourful desert-dweller is known as an
'earless' dragon because it is a member of a group of
lizards which have the ear opening or ear drum (one or
the other of which is visible on other lizards) hidden.

Right:
Ornate Dragon *Amphibolurus ornatus*
The Ornate Dragon is patterned to match the lichen-
speckled granite outcrops which it inhabits. Only on its
undersurface (where a reddish tone occurs) does it
depart from this concealing coloration. Long hind legs
and exceptionally long toes equip it to run at an
amazing speed across the open expanses of rock.
Occasionally it will stop to bob its head up and down, a
performance which may, by moving in relation to each
other various objects within the reptile's field of vision,
help it to gain a better impression of their size and
distances.

Above:
Western Netted Dragon   *Amphibolurus reticulatus*
Named for the 'reticulated' pattern of its colourful
scales, the small Western Netted Dragon is an
inhabitant of the dry scrublands of inland Australia.

Left:
Western Jew Lizard   *Amphibolurus barbatus minor*
This sub-species of the Bearded Dragon is smaller in
size and has a smaller "beard" of spiny scales.

Below:
Black-striped Snake   *Vermicella calonota*
The colourful little Black-striped Snake lives under
logs, in decayed wood or in termite mounds where it
feeds upon insects and other small creatures.

Right:
Little Whip Snake   *Denisonia gouldii*
The fast-moving but very small and quite harmless
Little Whip Snake is usually orange-red with a dark
brownish-black head; but other, usually uniformly
dark, varieties occur. This beneficial species is an
inhabitant of south-western Australia.

Left Above:

Bobtail *Trachydosaurus rugosus*
This richly coloured specimen of the Bobtail (also known as the Shingle-back Lizard) is from the laterite-capped Darling Ranges of south-western Australia. The reptile's colours exactly match the reddish rock and the patches of lichens. Bobtails from the sandy coastal plains only a few kilometres away are dull grey or grey-brown, while those from more distant parts of Australia are almost as diverse in coloration as are the rocks and soils of the continent.

Left Below:

Barking Gecko *Phyllurus milii*
This big-eyed Barking Gecko (so named for the little coughlike sounds it utters) pauses on mossy ground during its nocturnal hunting. If it sees an insect or other small creature it will probably first stalk it slowly, until within range for a sudden final leap forward. If its prey is large, a beetle, moth or mantis, there may follow a brief battle, with the insect being finally pounded into submission against the hard surface of a tree-trunk or rock.

Above:

Leaf-tailed Gecko *Phyllurus cornutus*
Australia's largest gecko, the Leaf-tail grows to a length of almost thirty centimetres. It is a nocturnal hunter which by day is perfectly camouflaged as it lies flat against a mossy tree-trunk.

Below:

Perentie *Varanus giganteus*
Largest of all the Australian goannas, the Perentie has a preference for the rocky ranges of the arid inland regions. From its retreat in the hills it may forage out into desert dune and scrub country. It is a very capable climber, and will investigate the stunted trees thoroughly for nests containing eggs or young birds, or any other prey; if disturbed away from its home among the rocks it will usually make for a tree as an alternative refuge.

Scale-footed Lizard *Pygopus lepidopodus*
Although rather snake-like in appearance, the Scale-footed lizard can be identified as being one of the legless lizards by the tiny scale-like vestiges of hind limbs, and by observing that the tail section is much longer than the rest of the body — snakes have a tail that is very much shorter than the body.

Carpet Snake *Morelia variegata*
Gliding down across a boulder a Carpet Snake shows the patterned back, like an old-fashioned carpet, from which its common name was derived. This python, which grows to a length of three to four metres, has the slit-shaped iris typical of the eyes of many nocturnal hunters.

Right:

Desert Pygmy Monitor *Varanus eremias*
Although looking rather like a more colourful but similarly patterned version of the common Sand Goanna *(Varanus gouldii)* the beautiful Desert Pygmy Monitor is much smaller, attaining a length of less than half a metre. Although dominantly bright yellow (often considerably dulled by dust) this goanna has fine spots and lines which form a rich colourful mosaic of black, white, yellow and rusty red. This patterning is most strongly defined down its back.

The termites very rapidly repair the damage, sealing in the eggs. As termites live in conditions of high humidity and warmth they provide ideal conditions for the incubation of the soft reptile eggs, and the very hard mounds give protection from predators. The young, on hatching, dig their way out.

Sand Goanna   *Varanus gouldii*

One of the most widely distributed of the big monitors of Australia, the Sand or Gould's Goanna occurs both in the arid interior and in wetter forested coastal regions. Because of this wide range of habitats it has a number of colour variants, and three distinct subspecies have been described. Typically it has prominent longitudinal markings from eye to neck; these may be a dark strip edged with pale yellow lines, or the dark area may be indistinguishable from a darker overall body colour, leaving only the yellow lines. The tail tip is always pale yellow. The Sand Goanna usually grows to a length of one and a half to two metres, and weighs five to seven kilograms.

Unlike the Perentie, *Varanus gouldii* rarely climbs trees, but travel very fast, sometimes rearing up and running on the two hinds legs. Its home is a burrow (often a rabbit warren) from which it forages for food, such as lizards, snakes and rabbits. Some of its food is obtained by digging.

THE LEATHERY TURTLE

FAMILY *DERMOCHELYIDAE*

This family contains but a single species, the Leathery Turtle or Luth, which differs from all others in having a unique carapace of small plates embedded in skin.

Sharp-snouted Flap-footed Lizard *Lialis burtonis*
The beautifully coloured and patterned, rather snake-like Sharp-snouted or Burton's Legless Lizard has an extremely wide distribution within Australia, and consequently occurs in many quite different patterns and tones, so that it could be thought that there were many rather than just a single species. The legless lizards are so snake-like that they are usually feared and killed whenever seen. Having no forelegs and only the smallest remnants of hindlegs these lizards move with serpentine sideways undulations, and when threatened pretend to strike in the same manner as a venomous snake.

Leatherback Turtle   *Dermochelys coriacea*

A huge sea turtle which grows to a length of about three metres and weighs up to 700 kilograms, the Leatherback is placed in a family of its own because, unlike any other turtle, it has a unique leathery carapace. With other turtles, the backbone is firmly attached to the carapace or shell, but in the case of the Leatherback the carapace consists of many small plates embedded in a tough leathery skin. This turtle, also known as the Luth, is the largest of all living turtles. It is common in Australian seas, but is thought not to breed in this region.

THE PYTHONS

FAMILY *BOIDAE*

The family Boidae contains not only the pythons, which inhabit Australia, Asia and Africa, but also the closely-related boas of the Americas. Most Australian pythons are nocturnal hunters, though some bask in the sun during the day. All are able to climb, and some species are exclusively arboreal in habits. Their prey is killed by the constricting action of the snake's steadily tightening coils; it is then swallowed whole, a feat made possible by the loosely articulated jaws and distendable skin. This family includes the largest of all Australian snakes (and one of the largest species in the world), the Amesthystine Python, of northern Queensland, which grows to a length of more than eight metres.

Children's Python   *Liasis childreni*

This widely distributed rock python is pale brown with darker uniform markings as cross bands or regular patches; a dark stripe on each side of the head passes through the eye. The Children's Python grows to a length of more than two metres, and closely resembles the Carpet Snake, but can always be positively separated from that species by the absence of any pitting on the rostral (broad scale forming the upper lip at the tip of the snout).

Like all pythons, this species is non-venomous; as it preys largely upon rabbits and rats it is beneficial, and deserves protection.

Green Tree Python
Chondropython viridis

The reptile fauna of Australia, particularly the north-eastern region, shows clearly the results of interchanges of species between Australia and New Guinea which occurred on a number of occasions when the sea level dropped thus linking these two islands, the most recent connection being between twelve and twenty thousand years ago.

One of the strictly rainforest-dwelling denizens of New Guinea which established itself in the jungles of Cape York was the Green Tree

Python; until 1940 this snake was thought to be restricted to New Guinea. This beautiful python, which is bright emerald green on its upper surface and yellow beneath, is a tree-dweller, where it coils itself into a tight bundle on a branch, with head poised on top. It grows to little more than a metre in length.

Rock Python   *Liasis olivaceus*

One of Australia's biggest pythons (and closely related to the larger Amesthystine Python), the Rock Python is generally restricted to the forested coastal parts of northern Australia. It grows to a length of four metres or more and is usually found in or near water waiting motionless for its prey, birds and mammals; a specimen now in the Western Australian museum has swallowed a half-grown kangaroo.

Scrub Python (Amesthystine Python)
Liasis amethystinus

Australia's largest snake is found in north-eastern Queensland, where it inhabits rainforests, eucalypt forests and rocky gorge country. Although the average length of Scrub Pythons is around five metres, some of between six and seven metres, and weighing more than 130 kilograms, have been found, placing this species among the largest snakes of the world.

Like all other pythons, it is not venomous, but kills the prey by enveloping it with the thick loops of the body, and contracting steadily to apply a tremendous crushing pressure. The victim is then swallowed whole.

Small mammals like bandicoots, probably form the major part of the diet, but on occasions Scrub Pythons have been known to capture, kill and swallow whole, such large prey as a Black-tailed Wallaby. Evidence of their stealthy approach and sudden seizure of prey is the finding of a Whistling Kite, probably caught at night, inside one Scrub Python.

Carpet Snake   *Morelia variegata*

This python is one of the largest in southern Australia, where it is very widely distributed. It is beautifully patterned in a manner resembling some old-fashioned carpets. The scales appear much smaller and more numerous than those of all the venomous snakes; there are more than forty rows of scales around the body, while the poisonous snakes have twenty-three or less rows around. The broad head is separated from the body by a distinct slender neck, while on the venomous snakes (except for the Death Adder) the head blends onto the body without any narrowing.

The teeth also of the Carpet Snake and other pythons are distinctive. Unlike the poisonous species which have fewer teeth, but have two larger grooved poison fangs, the pythons have a great many small solid teeth.

As the Carpet Snake preys upon mice and rabbits it is a beneficial species, and should never be killed.

SEA-SNAKES

FAMILY *HYDROPHIIDAE*

Sea-snakes are highly specialized reptiles with many adaptations that permit an exclusively aquatic life. They have of course remained dependent upon the atmosphere for their oxygen supply, and therefore must surface periodically to breathe; however their lungs permit very long periods underwater.

Sea-snakes are found throughout the tropical and sub-tropical seas of the Indo-Pacific region, mostly between India and the north coast of Australia.

Although their venom is extremely potent, their fangs are relatively short, and generally the sea-snakes do not bite unless molested; nevertheless many deaths have been recorded among the native fishermen of Malaysia and Indonesia.

Yellow-bellied Sea Snake   *Pelamis platurus*

The sea snakes have become highly adapted to the marine environment. Most have the body and tail laterally flattened, and the tail may be quite paddle-shaped. The nostrils are equipped with valves to close off the respiratory passages when the snake is submerged. Little is known of their habits, except that they eat fish and eels. About twenty species occur in Australian waters, mostly in tropical seas, but some in the temperate regions, occasionally reaching as far south as Tasmania.

One of the most common, often washed up on the beaches, is the Yellow-bellied Sea Snake. This species has an extremely wide distribution, ranging to the coasts of America and Africa. It is black above, cream to yellow below, and has a spotted or banded tail.

THE SKINKS

FAMILY *SCINCIDAE*

The great many skinks, about 150 species within Australia, are an extremely diverse group, and there are very few features common to all. Most are able to shed and replace their tails. Some lay eggs, but others bear live young; most have smooth glossy scales, but others have rough dull scales; some are heavily built with short thick stumpy tails; the majority are slender and long-tailed, while a few are legless, and snake-like in appearance. All species have scales that overlap, and the scales of the head are replaced by enlarged symmetrically arranged plates. While the smaller species are insect-eaters, many of the larger skinks accept a wide variety of foods, from flowers, foliage and fruit to insects and snails.

Blue-tongue Skink   *Tiliqua scincoides*

The common Blue-tongue grows to a length of fifty centimetres, and with a rounded bulky body, has legs that seem ridiculously small and inadequate. Most times the body rests on the ground, sliding on its smooth scales, 'rowed' along by the little legs. But when necessary the Blue-tongue can scuttle away at a surprising speed for a short distance.

The Blue-tongue occurs in all mainland States and Tasmania, and there are also several similar species and sub-species, such as the Western Blue-tongue and the Northern Blue-tongue.

When alarmed, the Blue-tongues gape their mouths and display their bright blue tongues; this display, accompanied by hissing, may deter some predators, but in actual fact the lizards are quite harmless.

The common Blue-tongued Skink, which like others of its genus gives birth to live young, is grey to brownish with darker cross bands.

Bobtail   *Trachydosaurus rugosus*

This large, slow-moving skink, which is readily identified by its rough pine-cone textured scales and stumpy fat tail, is also known as 'Sleepy Lizard', 'Stumpy-tail', and 'Two-headed Lizard'. It is widespread across southern Australia, and varies considerably in colour. For example, in south-western Australia it may be uniformly grey-brown, or chocolate brown with orange-red especially on the head, or orange-brown with creamy-white bands. A South Australian race is brown with blotches of bright yellow.

This is a sluggish reptile which cannot usually capture active prey. Vegetable materials — seedlings, flowers, fruits — as well as some insects and snails, are the principal foods. Cryptic colorations and armour-like scales are its principal protection.

Cunningham's Skink   *Ergenia cunninghami*

This large (thirty centimetre) heavily built lizard is one of the most common of the larger Australian skinks. Members of this world-wide family have hard, smooth and generally shiny scales; on this species the scales each terminate in a spine. Cunningham's Skink, also known as the Spiny Rock Lizard, lives in rocky country; when alarmed it hides in a crevice, where it puffs out its body so that the spines of the scales catch on the sides of the rock and grip very tightly. The tail is especially spiny.

This skink is blackish with small white spots. It lives upon insects, found among decayed wood of fallen trees or other debris.

Striped Skink   *Sphenomorphus lesueurii*

A small, fast-moving skink which inhabits open forest country and heathlands of mainland Australia, this species is neatly patterned in olive-brown with alternating longitudinal stripes of creamy-white and dark brown.

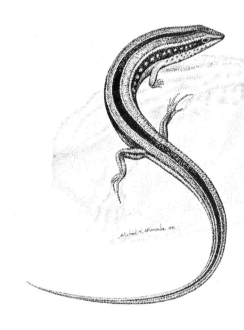

The skink family is a large one, with some 600 species, of which about one quarter occur in Australia. Most, like the Striped Skink, are diurnal. Typically, their scales are smooth, usually shiny, with larger, symmetrically arranged scales on the head.

Water Skink   *Sphenomorphus quoyii*

An active, alert skink of similar size to the Striped Skink, that is, about 20 centimetres in length. It is coppery brown above, speckled with black, and on the sides black mottled with white.

The skinks of the genus *Sphenomorphus* contain both egg-laying and live-bearing species; the Water Skink is one of those that produce live young.

This is the only Australian skink that is semi-aquatic, living at the edges of creeks, rivers and waterholes where it hunts insects. When disturbed it darts into the water, emerging cautiously some time later.

THE SNAKE-LIZARDS

FAMILY *PYGOPODIDAE*

The legless lizards of this family are restricted to the Australian region, two occurring in New Guinea and the remainder in continental Australia. Through the gradual evolutionary loss of the forelegs and the reduction of the hindlegs to a size where they are almost invisible, these lizards have very convincingly mimicked snakes both in appearance and behaviour, although they are of course completely harmless. The vestigial remains of the hind limbs are visible on some species as tiny scaly flaps, these are sometimes known as scaly-footed or flap-footed lizards. Although normally flat against the sides of the body, they are sometimes held out at right angles, and are then more easily visible. In spite of their very small size the flaps contain the almost microscopic vestiges of the bones found in a normal lizard hind leg, in some species complete even to the bones of the toes.

Scale-footed Lizard   *Pygopus lepidopodus*

Not only do these 'legless' lizards mimic snakes in appearance, but in behaviour also. When alarmed they raise their heads, flattening their necks as do many snakes when angry. If closely approached they may actually strike, as does a venomous snake, but they usually strike past their foe, and with mouth closed.

The colours of this widespread and quite common species are variable, being usually reddish, but ranging from light grey to dark brown, sometimes without the dark spots. It grows to a length of about seventy centimetres. Like other legless lizards of its family, it can shed its tail when this is necessary to escape an enemy.

Sharp-snouted Flap-footed Lizard
Lialis burtonis

Although appearing to be legless, and having a superficially snake-like appearance, this lizard will, on close examination, be seen to have vestigial hind limbs, tiny and flattened like little paddles, and normally lying flat, but occasionally held out at right angles to the body.

The Sharp-snouted or Burton's Flap-footed Lizard has successfully established itself in an extremely wide variety of habitats, from the deserts of the interior to the coastal forests; its distribution extends from southern Australia to New Guinea. Because it is so widespread there are many different colour and pattern variations, but always the species can be distinguished by its long slender snout.

Some individuals are grey with a 'herringbone' pattern, or longitudinal stripes. One form, from north Queensland, is golden yellow with dark stripes, while others again are brown, black or dark red.

When hunting, this lizard which is mainly nocturnal but may also be active at times during the day, is said to approach slowly, then when within striking distance, to lunge forward and downwards at its prey.

TORTOISES

FAMILY *CHELYIDAE*

All of Australia's tortoises (that is, the river and swamp inhabitants which have webbed and

clawed feet as distinct from the flipper-propelled ocean dwelling turtles) belong to this family. There are probably twelve species, which are divided into two groups — the long-necked and the short-necked tortoises. Although normally coming onto dry land only to lay their eggs, the tortoises occasionally travel considerable distances across country. Some of these journeys are forced migrations caused by drying up of rivers or swamps, but at other times, particularly in wet weather, they wander for no apparent reason. In their swamp and river homes the tortoises feed upon fish, crustacea, insects and some aquatic plants.

Krefft's Tortoise   *Emydura kreffti*

One of three eastern Australian short-necked tortoises, Krefft's tortoise is usually an inhabitant of rivers rather than swamps. It has distinctive yellow markings on the face, a feature shared by another, the Macquarie tortoise.

This tortoise is a carnivorous and scavenging species which is quite common in the north.

TURTLES

FAMILY *CHELONIIDAE*

Turtles, unlike tortoises, have all four limbs modified to form swimming flippers, enabling them to move swiftly and gracefully. In the water they are far superior to the tortoises which, however, have the advantage on land. Like the tortoises, the marine turtles have an armoured shell made up of a convex roof (the carapace) joined at the sides to an almost flat floor section (the plastron). The inner construction of the shell is of bony segments fused rigidly to the ribs and vertebrae, while the outer surface is of horny plates.

Green Turtle   *Chelonia mydas*

These big turtles, which may attain a total length of about two metres and weigh almost 200 kilograms, occur around Australia's tropical coasts from the Barrier Reef in the east to the Abrolhos Islands in the west. When small, Green Turtles are probably mainly carnivorous, but become more vegetarian, feeding on various seagrasses when larger. Most amazing is their fondness for the Portuguese Men-o'-war, which they are somehow able to eat without injury from the jellyfishes' stinging tentacles.

Green Turtles are most easily observed during their breeding season between November and March each year, when the females crawl ponderously up the sandy beaches of the reef islands to lay their eggs.

Instinctively the turtles return to the islands of their birth, and make long migrations to return to the few suitable places. The eggs, between fifty and 200 in number, are laid in a hole that the turtle digs with her hind flippers in the beach sand above the high tide level. Up to seven batches of eggs are laid by each female in one season, but only a very small proportion of the hatchlings survive the waiting seagulls, crabs, sharks and other predators when after ten weeks they emerge from the sand.

Loggerhead Turtle   *Caretta caretta*

A common turtle of northern Australian seas and the Barrier Reef, the Loggerhead does not appear to venture down cooler southern coasts. The name 'Loggerhead' results from the massive head, disproportionately large for the size of the body.

Large numbers of Loggerheads visit the Queensland coast and coastal islands, where the females dig into the sand to lay up to 100 eggs. When these hatch after about six weeks the young, each barely ten centimetres in length, dash to the sea.

Birds

Australia has about 775 species of birds, of which 600 breed on this continent or on nearby islands, and 125 are non-breeding visitors, mostly from the northern hemisphere. There are also about twenty species which have been accidently or intentionally introduced since the beginning of European settlement.

Among these are some of the world's most unusual and distinctive birds. Of the 600 breeding residents, some 360 are unique to this continent — an indication that there must be many that are quite unlike any birds of other lands. In recent years the theories put forward to explain the origins of this unique bird fauna have been modified in the light of new scientific discoveries.

It was formerly believed that Australia was colonized by birds from Asia, by a process of 'island-hopping', as birds of various species gradually found their way from island to island down the chain from Malaysia through Indonesia to New Guinea and finally into northern Australia.

The earliest arrivals were of course those that, now, have had the longest time in Australia. These are the endemic families which have been here and in New Guinea so long that their origins are obscure — they have no recognizable relatives elsewhere in the world. Later arrivals, which have had less time in Australia for evolution to proceed so far, have become endemic sub-families, endemic genera, species or merely sub-species. But it now seems probable that the origin of these oldest endemic families of truly Australian birds was not from Asia.

Until quite recently it was believed that the breaking up of the continents of the world occurred before the evolution and spread of birds, and therefore did not apply to theories of bird evolution and distribution.

But fossils of feathers from claystone strata in southern Victoria are probably of Lower Cretaceous age (about 130 million years) and could be considerably older. This recent discovery pushes the origins of Australian birds much further into the past, indicating that birds of some kind existed in Australia at a very early stage in the evolution of the birds of the world.

There is also evidence that the continents of Australia, Antarctica and Africa remained quite close together after the breakup of other parts of the great land mass of the original supercontinent. This would have allowed a sharing of many early bird types between these continents. At that time a vast expanse of ocean separated Australia from Asia.

So the older, unique, most unusual of Australian birds (those that were formerly thought to have been the earliest of the immigrants down the island chain from Asia) have probably been with this continent since it separated from the other southern continents. One group of birds almost certainly in this category would be the ratites — the Emu, cassowaries, kiwis and moas, all of which are accepted as having a common ancestry. Others of similar origin could be the megapodes, the Pied Goose, the scrub-birds; lyrebirds, parrots and bronze-cuckoos.

It is probable that Australia also had many other groups of birds which today are absent — the vultures, pheasants and flamingos (fossils of which have been found here). It was thought that these simply had not found their way down the island chain from the north, but the alternative explanation is that they were in Australia, but vanished due to climatic or other changes, or could not compete with competition from new arrivals from Asia.

It was not until much later that invading species began arriving from Asia, crossing one after another of the narrowing stretches of water, and finding suitable habitats within Australia. In so doing they must have displaced some of Australia's archaic birds, particularly those which had over-specialized and were unable to adapt very successfully to the great climatic changes which were occurring.

Within Australia, changes in the bird population resulted not only from these new arrivals, but also from the creation of new species by geographical isolation. Several factors were responsible.

The first was the upheaval which formed the great Dividing Range and split the fauna into two parts — a Bassian Fauna, inhabiting the forested wet coastal south-east and Tasmania, and an Eyrean Fauna across the remainder of the continent.

Secondly, the appearance and spread of the central Australian deserts cut the coastal fauna into pockets of varying size. Today the bird fauna is divided into a Bassian part (in south-eastern and south-western corners), an Eyrean part inhabiting the vast dry inland, and a Torresian part in the tropical north.

New Guinea, with its terrain of high mountains and deep isolated valleys, provided the isolated pockets of habitat that allowed the evolution of a great many new species, genera and families. Some of these spread to Australia (where they are mainly confined to the rainforests of the north-east) including the birds of paradise, crowned pigeons and cassowaries.

Australia has about three hundred and eighty endemic species of birds (including a few that have spread a little way into New Guinea or are occasional vagrants to that island). This is almost sixty per cent of the total of the species which breed here. The families which are found only in Australia include the scrub-birds, lyrebirds, chats and the Emu. Families which are almost entirely endemic are those of mudnest-builder, the butcherbirds, magpies and currawongs, sittellas, and tree-creepers.

The honeyeater family is well represented. Forty per cent of its species occur in Australia and of these, eighty per cent are endemic. The Australian parrots, although only seventeen per cent of the world total, are all endemic but for a few species. Other groups of birds strongly represented or unique to the Australian region are the fairy wrens, finches, flycatchers and robins.

Australia's many blue wrens, or fairy wrens, form a particularly colourful group. Although colour varies greatly from species to species, some having no blue at all in their plumage, their very long tails, held vertically most of the time, make these birds easy to recognize. The wrens occupy a wide range of habitats covering most of the continent: best known is the common Blue Wren of coastal south-eastern Australia.

The Australian robins belong to the Old World Flycatcher family, but unlike the typical flycatchers which form part of this family, the robins of Australia do not usually catch insects in the air. They sit motionless but for an occasional flick of a wing, on a perch or clinging to the side of a treetrunk, then drop to the ground to take their prey. There are five species with red or pink breasts, about six species with extensive areas of bright yellow, and a number of less colourful birds, including the White-breasted Robin, Dusky Robin, Hooded Robin and others.

Fantails fit the flycatcher image much better than the sedate robins, being active birds, always restless when perched, constantly fanning out their long tails, twisting this way and that, then suddenly fluttering up among the foliage, twisting and turning in complicated aerial acrobatics in pursuit of a fly or other small insect.

Others are of interest for their habits rather than colourful plumage. The bowerbirds and incubator birds in particular have fascinated ornithologists since their discovery, and research into the lives of these and many other Australian birds will continue to reveal fascinating new facets of their unique way of life.

ALBATROSSES

FAMILY *DIOMEDEIDAE*

The albatross family, together with the following three families, are ocean-wanderers which have in common a very characteristic feature — the nostrils are extended as tubes along the top of the bill.

Albatrosses are the largest of all seabirds; only the Giant Petrel is larger. They spend their entire lives at sea except when breeding, and have perfected the use of gliding to conserve energy by rising against the wind, then turning with the wind and swooping low over the wave-tops.

Shy Albatross *Diomedea cauta*

Albatrosses are, as a group, the largest of sea birds, rivalled only by the Giant Petrel. They have short stout bodies and extremely long narrow wings, which are extremely aerodynamically efficient. These birds excel at soaring just above the waves, gaining lift and speed, without flapping their wings, by utilizing differences of wind velocities from wavetops up to higher levels.

This species, also known as the White-capped Albatross, is a bird of the southern oceans; it breeds on islands in Bass Strait and off southern Tasmania.

AVOCETS AND STILTS

FAMILY *RECURVIROSTRIDAE*

All seven species of this family of graceful long-legged shorebirds have extremely long bills, but only the avocets have the bill recurved or bent upwards (giving the family name Recurvirostridae). These differences reflect the feeding habits of the two groups, the straight-billed stilts probing the mud beneath shallow water, and the avocets sweeping their curved bills from side to side as they walk through the shallows. The legs of stilts are longer, in proportion to their bodies, than any other bird except flamingoes. Their feet

are partially or mostly webbed, wings are long and pointed, but tails are short and square-cut. The plumage is usually boldly black-and-white, but some species have large patches of chestnut. One species of avocet and two stilts occur in Australia; two, the avocet and one of the stilts, are endemic.

Avocet *Recurvirostra novaehollandiae*

Of the world's four Avocets, one species occurs in Australia, where it is endemic. Avocets are characterized by extremely long upcurved bills, long legs, and webbed feet which have a very small hind toe.

The Australian Avocet is distributed throughout the continent except the far northern parts of Northern Territory; it will even be seen in desert regions after rain has replenished the normally dry salt lakes that are so common throughout much of inland Australia. When feeding the Avocet generally keeps to the shallows of lakes or coastal estuaries, walking steadily forward with probing and sideways sweeping actions of the long bill.

Avocets nest in colonies, the nests being slight hollows in the ground, sometimes built up a few centimetres with pieces of driftwood and other debris, and lined with soft vegetation.

Banded Stilt *Cladorhynchus leucocephalus*

The tall slender long-legged Banded Stilt resembles the Black-winged Stilt, but differs in having a broad chestnut band across the breast, and a dark brown line along the abdomen. The wings are brownish-black, and the head and neck entirely white.

The Banded Stilt inhabits salt lakes and brackish estuaries of southern Australia and Tasmania, nesting in colonies on the sandbanks of a few inland lakes that happen to suit its requirements. The nesting of these stilts, recorded on relatively few occasions, appears to be controlled by rainfall over the extensive salt lake systems of inland south-western Australia. The stilts gather in huge numbers on those lakes where conditions are best, that is, when the little brine shrimps upon which the birds feed, are most abundant. After breeding the big colonies disperse to the lakes of other parts of southern Australia.

Black-winged Stilt (Pied Stilt)
Himantopus himantopus

Crisply plumaged in pure white except for the black wings, black upper back, and a black patch on the nape of the neck, and with extremely long pink legs, this is one of the most graceful of our smaller water birds.

The Black-winged Stilt occurs over almost the entire continent except a few of the most mountainous and heavily forested regions.

Nesting is generally in colonies, and the speckled eggs may be laid in a small hollow in the ground at the edge of a lake, or a nest of water weeds and grasses may be built up like a little island, in shallow water.

This stilt, which is distributed around the world in temperate to tropical regions, has a sharp yapping call like the barking of a small dog.

BABBLERS

FAMILY *TIMALIIDAE*

Within Australia the true babblers are represented by the birds of the genus *Pomatostomus*, the Grey-crowned Babbler, Chestnut-crowned Babbler, White-browed Babbler and Hall Babbler. All are endemic except for a small population of the Grey-crowned in southern New Guinea. These babblers are dark grey-brown with white-tipped black tails, broad white eyebrow markings, and long slender down-curved bills. Their most distinctive feature is that they live throughout the year in groups, and are very noisily gregarious in their behavior.

Also in this family are other groups which are probably sufficiently closely related to babblers to be included in this family. These are the quail-thrushes of the genus *Cinclosoma*, the log-runners of the genus *Orthonyx*, and the whip-birds of the genus *Psophodes*. Like the babblers, they are all mainly terrestrial and are comparatively poor fliers. This widely distributed family has about 280 species, of which thirteen occur in Australia, eleven being found only in Australia.

Avocet *Recurvirostra novaehollandiae*
An Avocet, about to settle down on its clutch of five brownish, speckled and blotched eggs in a driftwood nest, shows the incredibly long upcurved bill and long slender legs of this species. In the background the brown-stained brackish waters of the shallow salt lake, rippling towards the shore, have through the camera's telephoto lens more the appearance of a stormy sky.

Chestnut Quail-thrush
Cinclosoma castanotum

The three species of quail-thrushes are terrestrial and, living in semi-desert country, have appropriate camouflage colours. The predominating ground colours of huge areas of the interior of Australia are the red-browns of clay and black, white and browns of rock. The Chestnut Quail-thrush is very colourful with the head and upper back brown, the lower back and rump chestnut, and the tail black tipped with white. The wing coverts are black tipped with white, and the flight quills brown. The throat and chest are black, with a white line extending back from the base of the bill. The belly is white, black around the thighs.

If disturbed the quail-thrush will run rather than fly, unless hard-pressed, when it will take off with a quail-like whirring of wings, fly a short distance then drop to the ground again. The nest is a small depression in the ground, usually under a bush or near a fallen branch. It is lined with strips of bark or grass; the two eggs are white spotted with brown.

Grey-crowned Babbler
Pomatostomus temporalis

Babblers, of which there are four species in Australia, are invariably seen in small parties, even nesting being undertaken by a group rather than a pair. Their activities are always accompanied by much noisy chattering. These flocks usually contain about six, sometimes as many as twelve individuals. They inhabit woodlands, including very dry country over the greater part of their range which extends from south eastern and northern Australia to the north-western coast. Here they feed mainly on the ground, flying up into the lower shrubs and trees every now and then.

The Grey-crowned Babbler is dark brown on the back and wings, with a chestnut band across the flight quills; the tail is very dark brown tipped with white. The crown of the head is grey-brown, which continues down to blend with the back. On each side of the head above the eye is a broad very pale grey eyebrow line. The throat is white, the breast chestnut, and the abdomen dark brown. The bill is long and down-curved.

The nest of the babblers is a unique structure. Although the birds are only about 25 centimetres in length a typical nest is about 60 centimetres high and 40 centimetres wide, and may be situated as high as 12 metres above the ground. As some nests are used for several seasons and new material added they may become much larger than this. The nest is con-

structed of quite large sticks, densely interwoven. The nest chamber is roofed over, and entered through a long horizontal spout near the top on one side. At times half a dozen such nests may be seen in one small tree. Often they are used (when not containing eggs or young) at night for roosting by the whole party of babblers. Two or three brown eggs are laid.

BARN OWLS

FAMILY *TYTONIDAE*

These owls have heart-shaped faces, the upper parts rufous-tinted and the underparts largely white with small dark flecks. The legs are long and feathered to the toes; the middle toe is equipped with a comb-like edge used for dressing the plumage. Anatomically, the barn owls differ enough from the other owls of the world to be placed in a separate family of their own. Like most owls, they remain paired for long periods, perhaps for life, and settle in the one locality, living in a large hollow, usually in a tree, or at times in cliffs or caves. Barn owls are known to be able to locate and capture their prey even in total darkness, using extremely accurate directional hearing to drop on any small mammal rustling among the leaves on the ground. Their plumage is extremely soft and silky giving almost silent flight. Barn owls are found throughout most tropical and temperate parts of the world; Australia has four of the total of ten species.

Masked Owl *Tyto novaehollandiae*

Closely resembling the well-known Barn Owl, with a large facial disc made up of feathers radiating outwards from around each eye, meeting between the eyes to form a ridge running up from beak to forehead. The Masked Owl is larger than the Barn Owl, darker and more strongly patterned on the back. The facial disc is white (brownish in the female) with a dark grey-brown border. The underparts vary from silky white with a few dark spots, to cinnamon with numerous grey and brown spots.

This owl inhabits forests and woodlands, usually in trees but sometimes in caves. The nest is a hollow in a tree or cave, where two white eggs are laid. It inhabits northern, eastern and south-western Australia.

BEE-EATERS

FAMILY *MEROPIDAE*

The bee-eaters are unmistakeable in appearance and behaviour. Plumages combine delicate pastel colours with strong bright hues. Wings are long and pointed, tails quite long but with the central pair of feathers narrow and greatly extended. The legs are very short, with the slender forward toes fused together. In flight bee-eaters are graceful and acrobatic, the wings vibrating

rapidly, then held rigidly out to glide; the birds wheel and dive like swallows in pursuit of flying insects, usually above the forest canopy or in open clearings. Bee-eaters are entirely arboreal except during the breeding season when they use their long bills to dig out a long tunnel terminating in a nest chamber. Some species nest in colonies. The young have their feathers protected against the dust of their underground home. They are hatched naked, but the feathers, when they appear, are wrapped tightly in waxy sheaths which finally split off to reveal plumage very similar to that of the adults; they remain in the tunnel until ready to fly. Twenty-five species occur throughout the warmer regions of the 'old world'. They are possibly of African origin, but have spread north to Europe, and eastward to Asia, and to Australia where only a single species occurs.

Rainbow Bird *Merops ornatus*

Conspicuous in southern Australian skies during summer months are the flashing bright orange wings of Rainbow Birds as they dart up from their perches on tree-top twigs or telegraph wires in erratic persuit of bees, march-flies, cicadas, dragonflies and wasps. These common but always impressively beautiful birds are regular summer migrants to southern Australia, but permanent residents in northern Australia and islands from Suva to the Solomon Islands.

Rainbow Birds breed during their stay in southern Australia, drilling a tunnel into flat ground or into a bank of soil. Up to five pure white eggs are laid in the unlined subterranean nest chamber.

BIRDS OF PARADISE

FAMILY *PARADISAEIDAE*

The birds of paradise are found only in Australia and New Guinea. The forty-three species range from starling size up to the bulk of a large raven. They are closely related both to the bowerbirds and the crows, with which they share a basic similarity of body shape. Plumages range from simple and unadorned, to the most elaborate plumes and ornamentations. Only males are so elaborately decorated, females being quite plain in appearance. The family is divided into a number of groups, including the manucodes (genus *Phonygammus*) and the riflebirds (*Ptiloris*) in Australia, and many true birds of paradise in New Guinea. The Australian Manucode is a glossy black bird rather like a Shining Starling or Spangled Drongo but with the long tail wide and rounded. The three species of riflebird are the nearest in Australia to the true birds of paradise; The Victoria Riflebird and the Magnificent Riflebird are unique to Australia, while the Paradise Riflebird occurs both in Australia and New Guinea.

Paradise Riflebird *Ptiloris paradiseus*

Riflebirds are inhabitants of the rainforests, where they probe the crevices of the bark with their long curved bills. The males are black; they

---

Rainbow Birds *Merops ornatus*
With an insect captured during a twisting, turning aerial chase, a fast-flying Rainbow Bird swoops upwards a fraction of a second before hitting the branch beside its mate. Its body is held vertically, so that its upward motion and the air-brake effect of its outspread wings and tail check its speed almost to zero at the last moment before landing; so perfectly is this done to reduce the impact, that the slender twig sways no more than a centimetre or two.

have short rounded wings, and short square-cut tails. They perform elaborate displays, using a bare horizontal limb as their stage. The male Paradise Riflebird is about 28 centimetres in length. His plumage, although black, has an iridescent purple-green sheen on the crown, centre of throat, front of the neck and central tail feathers. The plumage of the lower breast and abdomen are tipped with green, while those of the head and throat are short and scale-like. On the flanks the feathers are elongated to form plumes. The remainder of the plumage is black with a purplish gloss. The female is predominantly grey-brown, with buff and brown markings.

The riflebird's nest is well hidden in dense foliage high up in the rainforest. It is woven of vine tendrils and leaves, lined with softer fibres, and often decorated with the cast-off skin of a snake arranged around the rim. The two eggs are pinkish with streaks of brown.

BITTERNS, HERONS AND EGRETS

FAMILY *ARDEIDAE*

Within this family are two major groups of birds of considerably different appearance. In many respects they are almost opposites. The bitterns are shy and secretive denizens of the dense reedbeds, and are rarely seen; the herons and egrets, although wary, hunt on the open shallows of lakes and on wetlands completely devoid of concealment. The bitterns are largely nocturnal while the herons and egrets are diurnal. The bitterns react to danger by 'freezing' and so blending into the reeds; herons and egrets take to flight while an intruder is still at a very considerable distance. Herons and egrets have extremely long thin legs; bitterns have short legs. Herons and egrets are mostly plainly coloured in grey or white; bitterns have mottled and streaked plumage of buffy and brownish coloration. Herons and egrets are invariably seen from afar, but are usually silent except for an occasional croak while in flight; bitterns are more often heard than seen, making very loud booming calls which on a still night carry far from their reedy swamps and are possibly an origin of the aboriginal 'bunyip' legends.

Of a total of sixty-four species in this worldwide family, fifteen occur within Australia and one species is endemic.

Brown Bittern *Botaurus poiciloptilus*

The largest of Australia's four species of bitterns, the Brown stands about one metre tall. Like other bitterns, when disturbed in its marshy haunts it immediately 'freezes' into an upright position — the long neck is extended vertically and the long straight bill pointed at the sky. In this posture it blends remarkably well with the vertical stems of the surrounding reeds. This bittern is mottled in various tones of cinnamon, dark brown and buff, with blackish streaks, a colour pattern which blends well with the mass of dead leaves and black mud of the lower levels of a reedbed.

The Brown Bittern is nocturnal in habit, feeding on frogs, fish and crustacea; when moving from one swamp to another it probably flies at night. This species occurs in south-eastern and south-western Australia, and in Tasmania. The nest is a large platform of rushes and other vegetation.

Plumed Egret *Egretta intermedia*

Australia has three species of egrets, these being the Plumed Egret, the White Egret and the Little Egret. The Plumed Egret is intermediate in size, but as this difference is only slight, and as all three are entirely white-plumaged, identification in the field is not easy. It is further complicated by changes in colour of the bills from breeding to non-breeding seasons, the bill of the Plumed Egret being yellow or orange in the non-breeding season, and red in the breeding season, while the bill of the White Egret may be black, yellow, or black-and-yellow, and that of the Little Egret, black.

The Plumed Egret has long fine white nuptial plumes hanging down the back and

breast, but not on the nape of the neck, while the White Egret has plumes only on the back, and the Little Egret on the back, breast, and two long plumes on the nape of the neck.

The Plumed Egret is found through southern Asia, New Guinea, coastal northern and much of eastern Australia. It feeds in shallow water and swamps, and nests in colonies, generally in trees standing in water.

White-faced Heron *Ardea novaehollandiae*

Standing about sixty centimetres tall (with neck extended) the White-faced Heron is uniformly blue-grey, darker on the crown and main flight feathers, and lighter on the undersurfaces; the forehead, face and throat are white.

The White-faced Heron is a common and familiar bird of swamps, mud flats, roadside ditches, wet grasslands and the margins of tidal estuaries, where it feeds upon frogs, insects, small fish and crustaceae. It occurs throughout Australia, including the arid interior after heavy rains. To reach such areas it undertakes long nomadic flights, otherwise it is sedentary. The nest is an untidy platform of sticks in a tree, often high; the three to five eggs are pale green.

BOWERBIRDS

FAMILY *PTILONORHYNCHIDAE*

The bowerbirds are unique among the birds of the world in their skill as architects, their use of tools, and their use of colour in decorating their structures. There are eighteen species, nine occurring in Australia and seven unique to this continent. Most bowerbirds inhabit regions of mountain ranges and dense jungles, where observations are difficult and photographs rarely obtainable; but in Australia several species live in semi-desert regions. Bowerbirds vary considerably in size and shape; some are very brightly plumaged, others are plain. Their main feature in common is the bower. This is a specially prepared dancing stage or platform on the ground, or a decorated perch above the jungle floor. Some are simply clearings on the ground, while others are elaborate structures with walls or towers of sticks, decorated with a variety of natural treasures such as snail shells, colourful feathers, bleached bones, seeds, brightly coloured native fruits, flowers and mosses. The crowning achievement of bowerbirds is the mixing of natural pigments and the painting of the walls of the bowers. Two of these species use tools — crudely fashioned brushes of bark fibres or similar material. They thereby join the ranks of the few animals that deliberately use a tool to achieve a desired end.

Each bower is the focal point of a male's territory. He begins to build or repair it each year in the season when he comes to a sexually active state, and maintains it only through the four or five months of the breeding season. The females alone build the nest and attend the young.

The birds of one group in this family do not build bowers; both male and female of the catbirds together rear their young and behave much like ordinary birds.

Golden Bowerbird   *Prionodura newtoniana*

Unlike other Australian bowerbirds, which are known as 'avenue-builders', the small Golden Bowerbird constructs a 'maypole' bower, consisting of two towers of sticks up to two metres in height. The columns of twigs and other forest debris are built around two slender vertical saplings or vines usually about one metre apart, and connected by a horizontal perch where both sexes display. The towers and perch are extensively decorated with flowers, mosses and berries. The female alone builds the nest, a cup-shaped structure and cares for the young.

The male is bright orange-yellow on crown and nape, underparts, sides and tail, and dull golden-brown on the remainder of the plumage, while the female is dark olive above, grey below.

This bowerbird is confined to mountain rainforests of north-eastern Queensland, where it is quite a common species within its rather restricted habitat and range of distribution.

Green Catbird   *Ailuroedus crassirostris*

Australia has two species of catbirds, the Green and the Spotted. These birds are considered to be the most primitive members of the bowerbird family as they do not construct display bowers.

The Green Catbird, a rather plump bird of about 33 cm length, inhabits the rainforests of eastern Australia, but does not extend to the north-east which is the province of the Spotted Catbird. Its nest is an open cup-shaped structure, well hidden in a tangle of vines or other dense vegetation. The nest is attended by both parents, unlike the bower-building members of the family whose females alone raise the young.

Regent Bowerbird
Sericulus chrysocephalus

Of all Australian birds the Regent Bowerbird is one of the most spectacular in plumage colours. The female is unimpressive in brownish tones but the male (which is considerably smaller) is strikingly patterned in gold and black.

This bowerbird is an inhabitant of the rainforests of eastern Australia, where the male builds an avenue type of bower consisting of two parallel walls of sticks. This bower is small, roughly constructed and sparsely decorated when compared with the well-known structures of the Satin Bowerbird. The display given by the male to attract females to the bower is almost silent, unlike the noisy performance of the Satin.

The male takes no part in nest building or rearing of the young.

Satin Bowerbird
Ptilonorhynchus maculatus

One of the 'avenue-builders' — a term applied to those bower-birds which erect two parallel walls of sticks as a focal point for courtship displays and mating — the Satin is so different in its plumage colours and dexual dimorphesm that it is classified as a separate genus. It is found only in the dense forests, principally rainforests, of coastal south-eastern and north-eastern Australia.

The male bowerbird during spring months re-builds and decorates his bower usually situated in a small open space and performs short display flights above it, uttering harsh churring and grating calls. If a female arrives he parades around the bower, holding in his bill a favourite display object such as a colourful feather. Mating occurs when the female has been enticed between the walls of the bower. Many different females visit each bower during the nesting season.

Nesting duties, the building, incubation and rearing of the young, are all undertaken by the female alone, the nest being an ordinary cup-shaped structure in the foliage of a shrub or tree.

Spotted Catbird   *Ailuroedus melanotis*

In general appearance similar to the Green Catbird, but smaller (23 cm length); its head, face and chin are black, spotted with grey-brown, and its voice is rather less cat-like.

The Spotted Catbird is a race of the Spotted Catbird of New Guinea, and like the Green Catbird is a rainforest inhabitant, but confined to north-eastern Queensland and Cape York.

BUTCHERBIRDS, MAGPIES, CURRAWONGS

FAMILY   *CRACTICIDAE*

The ten species of this family are all inhabitants of the Australia-New Guinea region, nine occurring in Australia and seven being endemic. They are remarkable in their social behavior patterns, and for the fact that some species use tools to extend the capabilities of bills and feet. Butcherbirds, magpies and currawongs inhabit forests, woodlands and dry-country scrublands where most species prey upon small lizards, insects, small birds and their young, and various fruits. They have large heavily built straight bills with sharp and often hooked tips. Their heads are relatively large, the legs and feet strong. These birds feed mainly on the ground, but perch and nest in trees and are strong fliers. Plumages are mainly in clear-cut patterns of black and white. The butcher-birds are of particular interest for their habit of using tools to assist storage and tearing apart of the prey. The magpies are interesting in a different way. Some species live in social groups of about five up to twenty birds which strongly defend a large territory; the males breed with many of the females, which alone are responsible for nest building and incubation of the eggs.

Black-backed Magpie   *Gymnorhina tibicen*

One of three species of Australian magpies (the others being the White-backed Magpie and the Western Magpie), the Black-backed is principally black except for contrasting pure white areas around the back and sides of the neck, from shoulder to inner wings, and from rump onto tail. The tip of the tail is black, and the bill is white with a black tip.

This is a common and familiar bird over the greater part of Australia. The Black-backed Magpie inhabits open timbered country, dry woodlands and tree-lined rivers of the interior, and suburban parks and gardens. It is absent only from the northernmost parts of Queensland and the Northern Territory, from Tasmania, and from the south-west and south-east which are the provinces of the other two magpie species. The nest is a rather untidy but well built and lined large open cup-shaped structure built into the forks of slender branches. The three eggs are variable in colour, usually blue or brown with streaks and blotches of brown and dull purple.

Grey Butcherbird   *Cracticus torquatus*

The name 'butcherbird' originates from this bird's habit of impaling its prey, which may be frogs, fledgling birds, small lizards, or spiders, on sharp twigs or thorns. A Grey Butcherbird has been observed using the jagged end of a dead branch to grip the body of a nestling honeyeater to assist in tearing the prey apart. Often the bodies of lizards and other prey are left wedged into small forks of trees.

The song of the butcherbird is spirited and melodious, and may be heard at any time of the year.

The Grey Butcherbird, which is about 30 cm in length, inhabits open forest, woodland and heathland country of Australia except the far north and some of the very arid regions.

BUSTARDS

FAMILY *OTIDIDAE*

The bustards, of which there are twenty-two species widely distributed (especially in Africa) include some of the heaviest flying birds in the world. The single Australian species attains weights of up to fifteen kilograms, and has a wingspan of more than two metres. The bustards as a whole are shy birds which inhabit the grassy savannahs and semi-desert. They have long naked ostrich-like legs, and are three-toed. Although powerful in flight, and often travelling great distances on migrations, they prefer to walk, or if hard-pressed, to run away from danger, and take flight only as a last resort. This has made them easy targets, and very vulnerable. Their low reproductive rate does not help them to rapidly replenish their losses, and they have vanished from most areas except the most remote and uninhabited. Bustards have impressive courtship displays in which use is made by the male of its large inflatable throat pouch, to the accompaniment of a loud roaring sound.

Bustard   *Ardeotis australis*

This large bird stands about one metre high, and is solidly built. Its long thick neck and greyish-brown, patterned plumage blend effectively with the colour of the dry grasslands and savannah woodlands of its native haunts. It has a black head cap, and long white feathers hanging down over the lower breast.

The Bustard once occurred throughout Australia, but now is very rare in south-eastern and south-western more closely settled areas. Although protected by law, many are shot, so that this species is common only in the most remote areas.

BUTTON QUAILS AND PLAINS WANDERER

FAMILY *TURNICIDAE*

Very closely resembling the true quails, and favouring the same habitats, the button or bustard quails are in fact very different, and

belong not only to a separate family, but to an entirely different order, the Gruiformes, which includes cranes and other marsh birds. An obvious external difference is that they have three toes instead of four (the small hind toe being absent) and are unusual in having reversed male and female roles in some respects. The female is larger, it undertakes the courting in the breeding season, and leaves the male to incubate the eggs and rear the young. These quails and those of the preceding family have no doubt become closely alike by the process of convergent evolution, a result of similar habits and environment.

The Plains Wanderer is sometimes placed within this family as an arrangement of convenience. It has four toes like a true quail, has the reversed breeding behavioural roles of the button quails, has relatively long legs, plumage more like the bustards, and has a plover-like head and bill. This bird is now rare and in danger of becoming extinct as its natural habitat vanishes.

King Quail   *Coturnix chinensis*

There is between male and female King Quails a considerable difference in appearance. The male is very dark brown marked with blotches and streaks of black and buffy white; the throat is black, the breast dark blue-grey and the belly chestnut. The female, while similar to the male on the upper parts, has a white throat and the remainder of the underparts are buffy-white, barred with black. The Australian birds are a race of a species which is distributed northwards to India and China. Within Australia it inhabits the coastal parts of the Northern Territory, and in eastern Australia from Cape York to Adelaide, wherever the preferred habitat of grassy swamps and damp grasslands occurs. Here it feeds upon grass seeds and insects.

At times the King Quail will be encountered in small flocks; it is probably nomadic, wandering according to seasonal conditions. Breeding also appears to be determined by environmental factors, nests having been recorded in most months. The nest is a depression in the ground lined with grass and with grass stems bent over to form a roof. The six to eight eggs are spotted with brown.

Plains Wanderer   *Pedionomus torquatus*

A bird which is superficially quail-like, yet in many ways so unlike other quails that it may represent a separate family of its own, or alternatively may be considered as a subfamily of the buttonquails *(Turnicidae)* It has features in common with the true quails, the buttonquails, the bustard and the plover. The Plains Wanderer, as the name indicates, is an inhabitant of bare and open plains, not hiding among the grass to the same extent as the other quails. When disturbed it runs very fast and does not readily fly up.

The male, the smaller of the sexes, is grey-brown marked with white on the upper plumage, and mottled with black and buff on the neck and

wings. The throat is white, the breast very pale cinnamon marked with black and the belly buffy-white. The female has a white and black collar and a large patch of rufous brown on the upper breast. The nest is a shallow depression in the ground.

Stubble Quail   *Coturnix pectoralis*

The plump little Stubble Quail, which in general appearance has a very rounded body, short wings and tail, is boldly marked on the upper parts with black, brown and buff, and streaked with large arrow-like white markings. The male is light chestnut with black markings on the breast, and buffy-white on the abdomen, while on the female the throat is white, the breast has longitudinal dark brown and buffy streaks, and the abdomen is buff-white.

The name 'Stubble Quail' results from this bird's ready acceptance of farm paddocks, especially those with short grasses such as the stubble which remains after harvesting, as a suitable habitat in place of similar natural environments. The nest is a small hollow in the ground, lined with grass or other vegetation. Eight to fourteen eggs, yellowish with dark brown markings, are laid. This quail occurs over the entire eastern half of the continent, the south-west corner, and the coastal strip of the dry north-west.

CASSOWARIES

FAMILY *CASUARIIDAE*

Cassowaries are confined to the rainforests of New Guinea and tropical north-eastern Queensland. There are three species, of which one occurs both in Australia and New Guinea; two are confined to New Guinea and adjacent islands. In past ages the Dwarf Cassowary, now found only in New Guinea and other northern tropical islands, occurred in Australia as far south as New South Wales. Cassowaries evolved in times when rainforests were much more extensive. They developed large bony head shields (casques) which probably serve to push aside the tangled undergrowth; the wings are rudimentary, and the legs short (compared with those of Emu and Ostrich) but very heavily built. The bill is laterally compressed, the flight feathers are reduced to bare black spines, and the feet have three toes. The females are larger than the males. They have not fared well in competition with man, vanishing from regions wherever the natural vegetation has been removed or altered.

Plumed Egret   *Egretta intermedia*
Poised to strike with spear-like bill, a Plumed Egret watches intently some small fish or other creature below the surface of a tropical lily lagoon. The yellow bill distinguishes this species from the very similar Little Egret and White Egret, which have black bills. The Plumed Egret occurs on inland swamps and rivers, and, less commonly, on coastal waters of northern and eastern Australia.

Satin Bowerbird *Ptilonorhynchus maculatus*
A male Satin Bowerbird, his dark blue-black plumage
with a violet sheen reflecting the light, attends his
bower on the rainforest floor, and awaits the visit of
any female that may happen to be passing by, or be
attracted by his loud harsh creaking and hissing calls.

Regent Bowerbird *Sericulus chrysocephalus*
A male Regent Bowerbird finds a small insect or spider
on a mossy branch in the Queensland rainforest. Its
golden-yellow wings and deep-orange head and mantle
seem to glow against the satin-black of the rest of its
plumage. When this bird is seen in flight it seems that
its flashing golden wings trace a path of glowing colour
through the gloomy green caverns of the rainforest.
The female is far less impressive, being brownish with
a black patch on nape and throat.

Right:
Green Catbird *Ailuroedus crassirostris*
In Australian rainforests the strange cat-like calls of
the Green Catbirds are sometimes so frequently and so
regularly heard that they become accepted as
inevitable background sounds of the 'big scrubs'. Yet
seldom will these birds be seen, for they generally
keep to the upper levels of the forests, where their
plumage merges into the universal greenery of the rainforest

Cassowary   *Casuarius casuarius*

This very large flightless bird, about 1.5 to 1.75 metres tall, is confined to the rainforests of north-eastern Queensland; it also occurs in New Guinea and adjacent islands. Although normally shy and avoiding any encounter with man, if cornered or provoked the Cassowary can become fiercely aggressive and very dangerous. The heavy legs and feet are used in attack, the nail of the inner toe being elongated to form a deadly knife; a number of deaths have been recorded as a result of attacks by Cassowarys.

Wandering through the dense vegetation in pairs or small family parties, Cassowaries feed upon seeds, fruits and other vegetable material.

These solidly built birds are able to run through the jungle where a man can but slowly force his way through the tangled vegetation, at amazing speeds, possibly up to 48 kilometres per hour. The high bony head casque is used to deflect obstructions, while the coarse plumage protects and slips past the thorny vines.

Cassowarys begin to breed about July, and lay up to nine dark green eggs. The nest is no more than a roughly cleared space on the forest floor.

THE CHATS

FAMILY   *EPTHIANURIDAE*

Chats are small, rather plump terrestrial birds, most of which have adapted to arid conditions.

Orange Chat   *Epthianura aurifrons*
At its nest just a few centimetres above the mud on the edge of a desert salt-lake a richly coloured male Orange Chat, confronted with a choice of three, each seemingly desperate for food, hesitates a moment before thrusting the lot down the throat of one of them. That particular youngster will probably be a little less energetic in its hunger-motivated, head-wavering begging performance (which begins the instant the touch of the parent's feet shakes the nest) when a few minutes later the next meal is brought and one of the other young chats will get its turn.

Several species are common in the most desert-like regions, and show many behavioural adaptations to make possible the use of such a harsh environment. The Crimson, Orange and White-faced Chats are nomadic, wandering over great distances to those areas where conditions are most favourable, and breeding at any time of the year when sufficient rain has fallen to bring up the lush but temporary vegetation with its attendant abundance of insect life. The male chats are more brightly coloured or boldly patterned than the females. However in the case of the Desert Chat, both male and female have the same plumage (for which it has been placed in a separate genus). It is more solitary than other chats, which form flocks, and is apparently not nomadic. Not all chats have invaded the dry interior of the continent — the Yellow Chat depends upon a moist habitat of the marshy swamps and saltbush lagoons along the lower reaches of tropical northern rivers. This entire family of five species is unique to Australia.

Orange Chat   *Epthianura aurifrons*

Australia has five species of small, rather plump little birds known as chats; four belong to the genus *Epthianura*, and one (the Desert Chat) to the genus *Ashbyia*.

Three have orange or yellow on the breast and abdomen, these being the Orange Chat, which is deep orange with contrasting black throat and face, the Yellow Chat which is more yellow than orange, and the Desert Chat, which is a pale yellow and could be confused with the female of the Orange Chat. The remaining two species are quite distinct — the Crimson Chat and the White-faced Chat.

The Orange Chat occurs in desert and semi-arid areas, mostly well-inland, of all mainland States, frequenting the stunted sparse shrubs of saltbush and samphire country, generally around the margins of inland salt lakes where it finds insects on the ground.

CORMORANTS AND DARTER

FAMILY   *PHALACROCORACIDAE*

Adapted to a life on and beneath the waters of marine coasts and inland lakes, the birds of this family have very long sinuous necks, small heads, short legs and webbed feet. On land they are awkward, the legs being so far back (for better swimming propulsion) that they must stand or perch in an upright penguin-like posture. Their flight is strong, however, and they wander considerable distances to new feeding grounds. Cormorants and darters pursue their prey of fish underwater, and stay down for half a minute or more. Cormorants have long, slender bills with sharply hooked tips, while the darter's bill is narrow and spear-like. In Australia there are three black-and-white species of cormorants, two wholly black cormorants, and one species of darter — which sometimes is placed in a family of its own, the ANHINGIDAE.

Darter   *Anhinga rufa*

Although at first glance the Darter appears rather like a cormorant, there are numerous differences of appearance and behaviour. The Darter has an extremely elongated flexible neck (from which it gets its alternative name 'snake-bird') with a very small head which blends into the neck, and long straight bill. There is a special kink in the neck which by sudden straightening out propels the head and spear-like bill towards the fish or other aquatic prey. The Darter often swims with its body submerged, the long neck and head protruding like a periscope. Like the cormorants, the plumage of the Darter is not waterproof, so that they must spend a considerable time after swimming perched with wings outstretched waiting for the feathers to dry.

Yellow-faced Cormorant
Phalacrocorax varius

Australia has five species of cormorants of which three are pied and two entirely black; the Yellow-faced is one of the pied species, being black on the upper parts, and undersurfaces. The bare skin of the face is orange, becoming greenish-blue on the chin. The legs and the plumage of the thighs are black.

The Yellow-faced Cormorant occurs throughout Australia, but is absent from Tasmania. It feeds on the waters of estuaries, bays, lakes, coastal lagoons, large swamps and the more placid reaches of rivers, where individuals or huge flocks may be seen. Nesting is usually on islands in estuaries or lagoons, on the ground, in low bushes, or in trees such as mangroves. This species is also known as the Pied Cormorant.

Cormorants are diving birds with long snake-like necks, short legs and webbed feet; they are well equipped for a life in marine, river or lake environments. Australia has five species.

Four, the Black Cormorant, the Little Black, the Pied and the Little Pied, are able to perch and to nest in trees, and occur inland on rivers, lakes and swamps. The other species, the Black-faced Cormorant, is entirely of marine occurrence, feeding on sea and river estuaries, and nesting colonially on the bare rocks of coastal cliffs and islands.

The black-and-white plumaged Pied Cormorant occurs near coastal parts of Western Australia; in eastern Australia, where major rivers occur further inland, it may be seen many hundreds of kilometres from the sea.

CRANES

FAMILY *GRUIDAE*

The tall stately cranes occur almost throughout the temperate and tropical regions of the world; of the fourteen species one, the Brolga, is native to Australia, and occurs nowhere else. A second, the Sarus Crane, inhabits southern Asia and has recently been discovered in north-eastern Queensland; it may be a new natural addition to Australia's birds, or it may have been here a long time but overlooked because it is almost identical to the Brolga. All cranes have powerful voices, some of them among the loudest sounds produced by any birds, and produced in a modified windpipe which is greatly lengthened and convoluted. The voice is used in keeping flocks together, particularly on migratory flights, and during the dances performed by these birds, when they bow and leap, emitting loud trumpeting sounds. The famed dances of the Brolgas have their equivalent among other species of cranes.

Brolga *Grus rubicunda*

A slender grey bird with darker flight feathers and tail, and red on the head only (on the very similar Sarus Crane this red extends to the neck and face). The tall (up to 120 centimetre) Brolga is a stately bird, all its movements seeming deliberate, unhurried and graceful. This species is

well known for its 'dancing' displays, when a number of the birds leap and bow with a deliberate dance-like rhythm, the slow beating of extended wings imparting a floating motion to their movements.

Brolgas are usually seen on damp natural or cultivated grasslands, the margins of swamps and river floodplain pools, where they sometimes congregate in very large flocks. Their nests are on the ground in such places, and are rough platforms of sticks. The two dull white eggs are blotched with brown. This member of the crane family is confined to Australia (except for vagrants to southern New Guinea) where it inhabits much of the north and east of the continent.

CROWS AND RAVENS

FAMILY *CORVIDAE*

The crows and ravens are wily birds, notorious for their seemingly intelligent behavior which expresses itself in boldness, curiosity and cunning. They are among the largest of passerine birds, with massive more or less conical bills. The majority are omnivorous, including carrion, small and nestling birds, small reptiles, insects, seeds and fruits in their diet. The corvids are very adaptable and socially organized, and have been able to benefit from many of man's alterations to the environment. It is believed that these birds, especially the genus *Corvus* to which the Australian ravens and crows belong, represent a peak of evolution — if progress is measured in terms of mental capabilities rather than such criteria as size, ornamentation or elaborate display. There are five species of ravens and crows in Australia, of a world total of about one hundred; they are all very much alike, difficult to separate except by close study of feather details, behavior and voice characteristics.

Raven *Corvus coronoides*

This large, strong, heavily-beaked scavenger (and to some extent predator) is entirely glossy black, with blue-violet highlights. The concealed bases of the body feathers are dusky grey (on the very similar Crow they are white) and the throat feathers (hackles) are long and lanceolate on adult birds.

The Raven inhabits woodlands of south-western and most of eastern Australia except Cape York. The well-built stick nest is placed in a vertical fork high up in a tree; it is softly lined with bark, hair, or fur. The four or five eggs are pale green marked with spots and blotches of dark brown and olive brown.

As an example of the cunning of these birds, it is said that when a person approaches a Raven's nest, the sitting bird drops almost vertically to the ground on the opposite side of the trunk, then flies casually away as if quite unconcerned.

CUCKOOS AND COUCALS

FAMILY *CUCULIDAE*

Two of the main subdivisions of this family occur within Australia, the parasitic cuckoos, and the coucals which build nests and rear their young in a normal way. The typical true cuckoo is brownish above, and mottled grey and brown beneath, has a graduated tail, feet that have two toes forward and two backwards, and quite a large bill without a hooked tip. Some species have brighter colours, and many have patches of iridescence on the plumage. Many have a superficial resemblance to birds of prey. Most are predominantly insectivorous, solitary, and migratory. Young cuckoos hatched out by foster parents are somehow able to follow the migration routes earlier traversed by their parents. The coucals, which rear their own young, are larger birds with long heavy-looking tails. They inhabit dense vegetation on or near the ground, and are generally reluctant to fly. Twelve of the total of 127 species occur in Australia, three being endemic.

Pallid Cuckoo *Cuculus pallidus*

This large cuckoo with a long broad tail is distributed throughout Australia. Of its total length of thirty centimetres the tail accounts for fifteen centimetres. The male is a light, uniform brownish-grey on the upper parts, with the inner webs of the flight feathers spotted with white and the outer tail feathers toothed and barred with white. The underparts are light grey. The female differs in being abundantly spotted with chestnut and buff on the head, back and upper wing coverts, and the tail feathers have chestnut as well as white markings. The more spotted appearance of the female is readily apparent in the field.

The Pallid Cuckoo is a migratory species, but details of its patterns of movements are not well known. In south-eastern Australia the species is present from September to January, in the south-west from May to November, and in the far north (N.T.) from September to March.

In true cuckoo fashion the Pallid deposits its eggs in the nests of other birds, choosing those that construct open cup-shaped nests. Some host species are far smaller than this cuckoo (Spinebill and Tawny-crowned Honeyeaters) or of similar size (White-winged Triller, Magpielark). The egg is pale pink, sometimes with a few darker pink spots.

Pheasant Coucal *Centropus phasianus*

Although part of the cuckoo family, the coucals are distinguished from other cuckoos by the fact that they build nests and rear their own young. These are large birds, with a total length of around sixty centimetres, of which more than half is tail.

In breeding plumage the Pheasant Coucal (both male and female) is black, except for the wings which are bright chestnut variegated with buff and black barrings, and the tail which is black barred with buff and tipped with white. In the non-breeding season the plumage is dark brown on the crown to the upper back and streaked with white, the tail is barred buff, the wings are chestnut barred with buff and black, and the underparts are buff.

The Pheasant Coucal generally keeps to the denser vegetation, usually in damp places, rarely flying, and then with a laborious flutter and glide with the long tail streaming behind. The nest is near the ground, and is a globular structure of grass or leafy twigs, the entrance being an opening in one side. Three to five white eggs are laid.

CUCKOO-SHRIKES AND TRILLERS

FAMILY *CAMPEPHAGIDAE*

Many of the seventy species of this family have a superficial resemblance to shrikes or to cuckoos, but in actual fact have no close relationship to either family. Cuckoo-shrikes have the wings relatively long and pointed and the tail long and rounded. They are rather sombrely plumaged birds in grey, black and white. The bills of the birds of this family are rather shrike-like, being notched and slightly hooked. They feed upon insects, berries and other native fruits. Trillers are distinctively different from the cuckoo-shrikes, being smaller, and the males have bold black-and-white plumage patterns. Their generic and common names refer to their superb trilling songs. The members of this family are arboreal in habits, with the exception of Australia's unusual Ground Cuckoo-shrike, which feeds on the ground, but flies up into trees when disturbed, when resting, to roost, and to nest. Australia has seven species in this family, four cuckoo-shrikes and three trillers.

Black-faced Cuckoo-shrike
Coracina novaehollandiae

A slender, graceful bird with a long tail which accounts for almost half of its 33 centimetres total length. The general coloration is a delicate dove-grey, with the forehead, sides of the head, throat and upper breast a sharply outlined, strongly contrasting black. Also black are the wing quills and the tail, which has white on its very tip. Beneath, the lower abdomen and the under-tail are white.

The Black-faced Cuckoo-shrike is a common bird which occurs in forests (excluding the rain-forests and 'wet' eucalypt forests) and woodlands. It is to be seen around the suburbs of some of the cities and is easily identified by its peculiar habit of lifting one wing after the other just after landing. In South-eastern Australia this species appears to be nomadic; elsewhere it is sedentary. It feeds on insects, seeds and fruits.

The nest is very well concealed. It is a shallow cup-shaped structure neatly blended into the upper surface of a limb, often in a broad horizontal fork, where it scarcely interrupts the contours of the limb. It is made of bark and other materials densely matted together with spiders' webs. The two or three eggs are olive with spots of brown and red.

THE DRONGOS

FAMILY *DICRURIDAE*

Drongos are distinctive and unmistakeable. They have black plumages and iridescent colours, but most conspicuous are their long and sometimes ornamental tails. Other features are the stout bill with a sharp hook and subterminal notch on the upper mandible, and well developed rictal bristles (spiny feathers) at the base of the bill. Insects small and large make up their food; large insects are torn apart as by a bird of prey, one foot holding down the victim while the hooked bill is brought into play. Insects are attacked in the air, caught up in mid-flight or knocked to the ground. Most drongos are bold and aggressive birds, which drive off from areas they regard as their own territory, other much larger birds. Most of the twenty species are inhabitants of jungles and forests of Africa, southern Asia, the Philippines, and islands of the south-west Pacific; one species occurs in northern and eastern Australia.

Spangled Drongo *Dicrurus bracteatus*

The most distinctive feature of this bird is its long tail which curves outwards at its tip, and forks, forming a 'fishtail' shape; it is about 15 centimetres long, approximately half the total length of the bird. The Spangled Drongo's body is entirely glossy black, and with an iridescent green sheen on the wings and tail. On the head, neck and breast are iridescent blue-green spots (the 'spangles') and the eyes are deep red, very conspicuous set in the black plumage.

The Spangled Drongo frequents the margins of the dense and wet forests, especially rainforests and mangroves, where it dashes out from a convenient perch to capture a flying insect. This species breeds only in Australia — it occurs in New Guinea, but only as a non-breeding visitor.

The nest is a shallow cup-shaped structure of slender vine tendrils and fibres, woven into a thin horizontal fork of a tree. The three or four eggs are pale pink with red and purplish spots.

EMUS

FAMILY *DROMAIIDAE*

Like the Cassowary family, the emu has three toes on its massive feet; its bill however is laterally broad rather than compressed, and it lacks the head casque. It inhabits a totally different environment of open grasslands, savannah woodlands and dry semi-desert scrublands. In spite of these differences, the emu family has many similarities with the cassowary family, and may have evolved from the early cassowaries, the changes since that time reflecting the vastly different habitats chosen by the two families.

There is only one species in the family (races of this species formerly inhabiting Tasmania have become extinct) and the family is endemic to Australia.

Emu *Dromaius novaehollandiae*

Second-largest bird in the world after the Ostrich, the Emu stands two metres tall and measures two metres from head to tail. It is flightless, but capable of running at speeds up to 48 kilometres per hour.

Once found throughout Australia except for the north-eastern rainforests, the Emu has remained quite common, abundant in some areas, probably because it inhabits desert and dry scrub country far from the centres of human population. It has, however, been exterminated from Tasmania and Kangaroo Island, and has vanished from most agricultural regions, where great numbers have been killed.

Emus are sedentary while conditions are favourable, but become nomadic and wander considerable distances to avoid food shortages or water, as at times of drought.

Breeding commences in autumn, when six to eleven very large dark grey-green eggs are laid.

FALCONS

FAMILY *FALCONIDAE*

The falcons include some of the most aggressive of all birds, representing a peak of evolution among the birds of prey. They take their prey, generally other birds, by pursuit, using superior pace and aerial agility, or by diving ('stooping') from above at a tremendous speed, said to be as great as 190 kilometres per hour; the prey is seized in the talons or stunned with a blow of the closed foot. There are exceptions — the small Kestrel and Brown Falcon capture their prey (small mammals and large insects) on the ground, by hovering or swooping down from a high vantage point. The typical falcon, such as the Peregrine, is powerfully built, with smooth contours and narrow pointed swept-back wings; not only is it capable of very fast flight, but from ts appearance alone gives the impression that it has the potential of far greater speeds. Of the total of sixty-one species, six occur in Australia, and two (the Grey Falcon and the Black Falcon) are confined to Australia.

Little Falcon *Falco longipennis*

Almost a smaller edition of the cosmopolitan Peregrine Falcon, this species also is distributed throughout Australia, but favours the open and lightly timbered country rather than the ranges. It builds a nest of sticks, lined with leaves, in a high fork of a tree rather than use a cliff ledge or tree hollow. The Little Falcon can be differentiated from the Peregrine by the absence of the black bars across the belly, and the underwings are barred with brown or buff instead of black; it has a length of thirty to thirty-five centimetres compared with thirty-eight to forty-five centimetres for the Peregrine (in both species the females are the larger).

In hunting technique also the Little Falcon resembles its larger relative, diving upon the prey or catching the bird or large insect by use of superior speed and manoeuvrability; at times the victim is larger than the hunter itself.

Kestrel *Falco cenchroides*

The little Kestrel is a familiar sight in many parts of Australia. Cinnamon brown on the back, buffy-white beneath, and black patches below the eyes, the Kestrel is a handsome small falcon which is best known for its hovering ability. When hunting for its usual prey of insects, (especially grasshoppers), spiders, lizards and other small prey, the Kestrel may drop onto its victim from a vantage point on tree or pole. But it very often hovers, maintaining a stationary position with great skill, then suddenly folding back its wings, it plummets into the grass.

The Kestrel, which is found throughout Australia except in heavily forested areas, usually selects a large hollow in a tree or a ledge or hole in a cliff as its nest, but occasionally will use an old nest of another bird. The four or five white or buff eggs are densely streaked and blotched with reddish brown.

Peregrine Falcon *Falco peregrinus*

In action one of the most thrilling of all the birds of prey, and famed for its 'stoop', when at tremendous speed it dives almost vertically downwards like a projectile, wings half closed, to strike its prey a killing blow with its powerfully built feet. The victim is almost invariably a bird, often as large as the falcon itself.

The Peregrine Falcon is dark blue-grey on the upper surfaces, black on the crown of the head and the face, and buff or rufous below, with bold black barring across the belly. It occurs throughout Australia and Tasmania (and almost world wide), but is more likely to be seen in mountainous country, where its nests on the ledges of cliffs. At other times it will use large hollow of trees, or the old nests of other birds.

FLOWERPECKERS

FAMILY *DICAEIDAE*

These are small birds of the forest crown, where they are generally solitary or in pairs rather than flocks. All of the flowerpeckers are fond of sticky berries and seeds, which are swallowed whole; later they are vented with some of the sticky layer still adhering, and are likely to stick to the high branches of trees. As mistletoe berries are common in their diet, the seeds are thus placed in position to begin parasitic attachments to the trees. The true flowerpeckers construct small purse-like pendant nests entered through a slit-like opening near the top. The nests are woven

and felted of spiders' webs, cocoons, fluffy seeds and other soft materials. However the diamond-birds of Australia are in many ways unlike typical flowerpeckers, which take tiny insects from foliage, flowers and bark, and have greatly departed from the flowerpecker pattern of nest construction — they build domed nests in hollows of trees or tunnels in the ground. Flowerpeckers are considered to be most closely related to the sunbirds, and many species are brightly coloured. The family is distributed from India and south-east Asia to Australia, where nine of the total fifty-eight species occur, eight of these being endemic.

Mistletoe Bird *Dicaeum hirundinaceum*

Found throughout Australia, even in near-desert regions, and extending northwards to the Indonesian and Aru Islands, but not south to Tasmania, the Mistletoe Bird has found great success in its association with the widespread parasitic mistletoe plants. These small, rather robin-like, scarlet-and-black birds feed largely upon the berries of the many species of mistletoe which commonly grow upon the branches of eucalypts, acacias and other trees; in doing so they distribute the mistletoe seeds to new host trees. Insects are included in the diet, and young birds in the nest are fed mainly upon insects while still very small. The female is much less colourful, mostly grey and brown, with just a touch of pale red on the under tail coverts.

FINCHES

FAMILY *ESTRILDIDAE*

This family contains a large number of small colourful seed-eaters which although similar to the European finches, are much more closely related to the weaver-birds which are common in Africa. (Included in this family along with the finches are the waxbills, mannakins and Java sparrows which do not occur in Australia). The finches are small, mostly less than twelve centimetres, and have short powerful bills. Their plumages are boldly patterned and colourful; some, like the Gouldian Finch and Red-eared Firetail Finch, are among the most beautiful of small Australian birds. The majority of species inhabit savannah grasslands of tropical northern and the interior of Australia. Of the one hundred and fifteen species in this family Australia has nineteen, of which fourteen are endemic.

Diamond Firetail *Zonaeginthus guttatus*

The Australian finches include many colourful species, most of which are birds of the grasslands of the tropical north and northern inland areas. But three species, known as fire-tails for their brilliant red rumps, are rather unusual in that they inhabit forest country.

The Diamond Firetail has a distribution extending down the east coast, from about Cardwell in north Queensland, to Victoria, and it occurs in coastal South Australia as far west as the Eyre Peninsula.

The colours of the boldly patterned plumage are (apart from the crimson area above the tail) a crimson ring around the eyes, a broad black band across the breast, black flanks spotted white, and dull grey-green back and wings. There is a black patch between the crimson eye ring and the crimson bill, while the throat, abdomen and under tail covets are white.

Like most other finches the Diamond Firetail is a sociable species, gathering into small flocks which search on the ground for grass-seeds. The nest is a large globular grass structure within a long narrow entrance neck.

FLYCATCHERS AND ROBINS

FAMILY *MUSCICAPIDAE*

The flycatchers feed almost entirely upon insects captured on the wing, so that they have well developed powers of flight, though not as highly evolved as other families such as the swifts. They generally have small bills with wide gapes achieved by a flattening of the base of the bill; their upper mandibles are fringed with long rake-like rictal bristles which serve to extend the catching area and increase their success in snapping up erratically flying insects. Since the legs and feet are used only for perching they are generally quite frail. Their wings have ten primary feathers, and tails are often long and may often be fanned out or wagged from side to side. Most Australian species are boldly patterned, often with bright colours. The flycatchers generally build open cup-shaped nests. There is some uncertainty as to the limits of this family — whether certain groups should be included or omitted; Australia probably has twenty-three of the total 320 species, and 13 of these are endemic. General included are the flycatchers of Australia and New Guinea (genus *Microeca)*, the five groups of red-breasted, yellow-breasted and

---

Emu *Dromaius novaehollandiae*
As one Emu drinks its mate remains warily on guard. Adult Emus have shaggy dark grey-brown loose plumage of long coarse filamentary feathers; the chicks are much more attractive with light and dark longitudinal stripes.

other robins (*Petroica, Eopsaltria, Peneoenanthe, Heteromyias,* and *Poecilodryas),* and the fantails (*Rhipidura*).

Hooded Robin   *Petroica cucullata*

In its general appearance this bird has the look of a typical robin, but it has neither the red of the red-breasted group of robins nor the yellow of the yellow-breasted group — it is entirely black-and-white. The dominating colour is black, including the entire head, the back, terminal half of the tail, and the wings. It has white on its breast, abdomen, and shoulders, and has a white wing stripe.

The Hooded Robin inhabitats open forest and dry inland plains across most of the continent, except parts of the tropical north; it is absent from Tasmania. In habits it is similar to other robins, using low trees, shrubs and logs as vantage points from which to rise into the air or drop to the ground to take an insect. The nest is a neat cup made of grasses or bark matted with spider webs and lined with softer materials; it is usually within several metres of the ground.

Mistletoe Bird   *Dicaeum hirundinaceum*
A male Mistletoe Bird speeds vertically upwards to his unique suspended nest which is constructed entirely of fine plant fibres, spider egg sacs, fur and other soft materials matted together with spider webs until resembling knitted wool in texture. The entrance is a narrow slit in the side which almost closes except when a bird is entering or leaving — or when three hungry young are reaching out with wide open beaks for a feed.

Pale-yellow Robin   *Eopsaltria capito*
A Pale-yellow Robin touches down on the rim of its nest in the rainforest. As well as being hidden by fern fronds and mosses the nest itself is camouflaged with pieces of lichens and moss bound on with cobwebs. More often than the vertical sapling fork chosen as a nest site by this pair, a prickly lawyer vine will be used. The Pale-yellow Robin, a quiet small bird of the jungle, is easily overlooked unless one reveals its presence by movement, as it drops to the ground to take an insect.

Red-capped Robin   *Petroica goodenovii*
Although his feet have already gripped the twig this male Red-capped Robin's wings (one concealed behind his head) are still outstretched in the action of 'applying the brakes' for a soft landing. The well camouflaged nest, originally very neatly constructed, is becoming tattered. The young robins, which will make their first flight within a few days, scramble forward with necks outstretched and beaks gaping for each meal that arrives.

Pale-yellow Robin   *Eopsaltria capito*

One of four Australian species of yellow-breasted robins, the Pale-yellow Robin is only slightly paler than the other species, but more a greenish yellow; the upper parts are olive green and the face is conspiciously marked with white which extends to the throat.

The distribution of the Pale-yellow Robin is of interest in that there are two distinct populations separated by well over a thousand kilometres of unsuitable habitat which has prevented interbreeding, and caused the evolution of two distinct races. The birds from northern New South Wales and southern Queensland (colour plate, the race *Capito)* are slightly larger and pure white on the face and eyes, while those of north-eastern Queensland have the white face areas tinged with rufous. These Yellow robins are confined to rainforest habitats, where they show a preference for areas thick with spiny lawyer vines, which are commonly used as for nest sites.

Red-capped Robin   *Petroica goodenovii*

The most widespread of Australia's five species of red-breasted robins (the others being the Scarlet, the Flame, the Rose and the Pink) the Red-capped Robin inhabits the entire continent except for high rainfall and generally coastal areas of the tropical north-east, south-east and south-west; it is absent from Tasmania. This is the only species with a red forehead patch. The female is conspiciously different, being grey-brown above and dull white below, and having only a dull rusty-brown tinge on the forehead.

In the more arid inland regions subject to drought this robin is nomadic, wandering wherever conditions are best; in southern parts its movements follow a regular seasonal migration pattern.

Like other robins, the Red-capped hunts from a low perch, dropping to the ground every now and then to take some insect or very small lizard. During spring months the pleasant ticking calls of this robin are among the commonest of bush sounds thoughout Australia's vast areas of mulga and mallee scrub.

Rufous Fantail   *Rhipidura rufifrons*

One of the daintiest most active of forest birds, always fluttering about, sometimes near the ground and sometimes high in the treetop canopy. Often, as it twists restlessly on a perch or works over the bark of a treetrunk or fallen log, the long tail is fanned out, showing its bright colours. Of the bird's total length of 15 centimetres the tail is more than half, about nine centimetres. The back is brown, lightening on the rump to reddish-brown, and becoming bright rufous on the upper tail coverts and orange on the basal half of the tail; the remainder of the tail is dark grey or grey-brown, with a white tip to each feather. The forehead and a line extending back over the eye are rufous; the face around the eye is almost black, the throat is white, and the upper breast is black becoming more and more broken with patches of white until buffy-white

on the belly. The flanks and under tail-coverts are cinnamon.

The Rufous Fantail inhabits the forests (especially rainforests) of coastal northern, eastern and south-eastern Australia. Its nest is small, cup-shaped with a slender tail hanging below, and very neatly constructed of bark, grass and cobwebs. The two or three white eggs are white spotted with pale amber and brown.

FRIGATEBIRDS

FAMILY *FREGATIDAE*

Although named for their piratical way of life, robbing other sea birds of their catch, the frigatebirds also find for themselves a substantial part of their food, usually snatching it from the surface of the sea. Their very long wings, long forked tails and very reduced legs reflect their airborne life. So degenerate are the legs that frigatebirds must land where they can easily launch out into the air, as from a tree or cliff. The bills are long and hooked. The plumage is usually predominantly black, but the males have a colourful throat patch, which can be inflated during the breeding season. Of the total of six species, two occur near Australian shores and one breeds on the coastal islands.

Greater Frigatebird   *Fregata minor*

Frigatebirds are large gliding and soaring birds of the tropical seas, so adapted to a life far out to sea that they even feed and drink on the wing. They never alight on the water, but snap up flying fish which they take in mid-air, and any small fish or other marine creatures that swim or

float on the surface of the sea. Frigatebirds bully and harry other sea-birds, forcing them to give up their food to their attacker, and preying upon their young; they are the pirates of the oceans and coasts.

The Greater Frigatebird (like the one other species of Australian seas, the Lesser Frigatebird) has a small body in relation to its huge wings and large forked tail. The legs and feet are small and weak, while the bill is long, heavily built, and hooked. During courtship the male inflates his crimson throat pouch, which when fully distended becomes like a huge red balloon.

The adult male Greater Frigatebird is entirely black except for the light brown wing-coverts and red throat; the female is white on the breast, while immature birds are white below, with a russet patch on the breast.

Frigatebirds nest in colonies, building a stick nest on a bush, and laying just one egg.

FROGMOUTHS

FAMILY *PODARGIDAE*

The twelve species of this family are rather owl-like, with soft mottled plumage, but with an enormously wide, short, hooked bill. Various colour phases occur, but in any one colour phase male and female are alike. All species have powder downs (which fray to provide fine powder for maintaining plumage condition) but lack the oil gland; there are ten tail feathers. These are rather lethargic birds, nocturnal in habits, hunting mostly on the ground. They are inhabitants of the open woodlands and tropical forested savannahs of the southern Asian and Australian regions. The family is divided into two genera, of which only one occurs in Australia; however all three *Podargus* species, the Tawny, Papuan and Marbled Frogmouths, occur here. Like owls, these birds have very large eyes, but of striking colours — the Tawny has bright orange-yellow eyes, the Papuan red eyes, and the Marbled Frogmouth, orange eyes.

Papuan Frogmouth   *Podargus papuensis*

This larger version of the Tawny Frogmouth occurs in north-eastern Queensland, and, as the name indicates, also in Papua. It is rather more spotted and marked with white than the Tawny, and its eyes are red instead of yellow.

The Papuan Frogmouth inhabits forests and woodlands, and in habits is probably similar to the Tawny Frogmouth. The nest is a rough stick structure placed in a fork of a tree. One white egg is laid.

Tawny Frogmouth   *Podargus strigoides*

The Tawny Frogmouth is an aberrant and exceptionally large form of nightjar, which at night hunts for insects, small reptiles and probably mouse-sized mammals which it captures on the ground. During the day it sleeps on a tree branch; if disturbed it will 'freeze' into an upright posture along the branch. Because of the mottled browns and greys of its plumage, the bird resembles very closely a short broken branch or one of the dead grey hollow spouts so common on eucalypts. While holding this absolutely motionless posture the Frogmouth watches the intruder with its very large yellow eyes narrowed to thin horizontal slits. If disturbed at night the bird does not adopt this broken branch attitude.

The Tawny Frogmouth is patterned with grey and brown giving a marbled or textured effect similar to weathered grey wood. The undersurface is a pale streaked pattern. The bill is exceptionally wide, and there is a tuft of bristle-like feathers at its base; this cavernous mouth presumably makes a most efficient insect trap at night. The nest is a very untidy (and small for the size of the bird) platform of sticks in a horizontal fork of a tree.

GANNETS AND BOOBIES

FAMILY *SULIDAE*

These are large sea birds which fish by plunge-diving from considerable heights. They have large bills which have only a slight terminal hook, and otherwise are straight and taper outwards to blend onto the head. As a modification for their diving habit, which results in their hitting the water with great force, they lack the usual nostrils into which water could be forced on impact. Instead, the mouth is extended back behind the eye and the extra section is covered by a section of the hard bill, which remains open to permit breathing. Gannets belong to the genus *Morus*, and have only a narrow dark naked throat stripe; boobies, the genus *Sula*, have extensive coloured facial and throat areas of naked skin. Four of the total of nine species occur in Australian seas.

Gannet   *Morus serrator*

Australia's single species of gannet is quite a large bird of about 1.75 metres wingspan; it is white except for the black wing tips, black trailing edge of the outer half of each wing, and black central tail feathers. The head and neck are yellow, and the bill blue-grey.

Gannets breed in colonies (known as 'gannetries') usually above cliffs on an island, or other place of relatively difficult access. The nests are hard mounds of sticks, earth and excreta, with a small depression on top where the one egg is placed.

Feeding is done at sea, and is often a co-operative effort by a flock. The gannets dive vertically into the sea from heights of twenty metres or more, throwing up a column of water as they penetrate the surface with great force and speed that carries them deep down towards their prey of fish or squid.

Red-footed Booby   *Sula sula*

Somewhat smaller and more slender than other Australian boobies (the Brown Booby and the Masked Booby) the Red-footed has a wing span of around one and a half metres. The plumage patterns of this species vary greatly, there being two main colour groups, but also many intermediate forms. A *white phase* has head and neck with a light buff or golden tones; the primary, secondary, primary coverts and major secondary coverts black, and the remainder of the plumage white. A *grey-brown phase* is buff to golden on head and neck, brown on the back, white on the tail, and elsewhere ashy-grey or grey-brown, bare skin around the eye blue and the legs red or pink.

The Red-footed Booby inhabits tropical seas, and is most common to the north-east of Australia. Nesting occurs on Raine Island, off Cape York, and islands of the Coral Sea. It feeds like the Gannet, by diving vertically from considerable heights to take squid and fish.

GREBES

FAMILY *PODICIPEDIDAE*

So adapted and dependent upon water that they can hardly walk on land, and if by mistake landing at night on land are unable, because of their weak legs, to take off again, the grebes compensate with superb swimming ability. The legs are placed very far back on the body for more efficient propulsion by the feet, which are not webbed but have each toe widely fringed to form paddle-like lobes. Out of the water they are poorly balanced, and walk very clumsily, so they rarely leave the water except to climb onto their floating nests.

Grebes swim well, and are expert divers, submerging with scarcely a ripple, staying long underwater and able to travel a considerable distance. One species, the Hoary-headed Grebe, is endemic to Australia.

Crested Grebe   *Podiceps cristatus*

The Crested Grebe is a striking large water-bird that could hardly be mistaken for any other. On the crown of its head is a double black crest, while in the breeding season a spectacular adornment develops. This erect ruff around the back of the cheeks consists of elongated feathers, chestnut at the base and black at the tips.

The Crested Grebe frequents open areas of water such as lakes, bays and lagoons. It builds a large floating platform of rushes and other aquatic plants, and incorporates a quantity of mud to stabilize the whole structure.

Although very widely distributed through Africa, Eurasia, south-east Asia, Australia and New Zealand, the Crested Grebe is an uncommon species, but likely to be found wherever there is suitable habitat.

GULLS, NODDIES AND TERNS

FAMILY *LARIDAE*

Gulls are gregarious scavengers of the seas and coastal rivers, feeding on shore debris, floating scraps of food, and occasional fish. The gulls are graceful long-winged sea birds of almost worldwide distribution. Gulls are the more stoutly built; the terns slender and dainty, with rapidly beating wings and erratic flight. Gulls have slightly hooked bills, the terns' are straight-pointed; gulls have their tails rounded, the terns, forked. Gulls fly with the bill pointed straight forward, the terns with the bill directed downwards. Gulls generally alight on water or land to feed, the terns dive for live prey. However between some of the terns and some of the smaller gulls, the differences are slight. The noddies are a small group which link other gull and tern groups. These birds are almost always in large flocks, and nest in large and noisy colonies. Of the total of about eighty species of gulls and terns, twenty-four occur in Australia, but only one is endemic.

Caspian Tern   *Hydroprogne caspia*

This very large tern can immediately be distinguished from all others in Australia not only by its huge size, for a tern (length about fifty centimetres) but also by its massive blood-red bill and

plumage which is almost entirely white except for the black head cap.

The Caspian Tern hovers well above the sea, head pointed downwards, until it sights a fish, when it plunges vertically, beak-first into the water.

A bird of the shallower coastal waters, inland lakes and large rivers, the Caspian occurs

as a breeding species around the entire Australian coastline and in suitable habitats inland. Beyond Australia it has a wide distribution including North America, Africa, European Asia and New Zealand.

The nest is merely a slight hollow in the ground. The one or two eggs are oval, rough-textured, stone-grey or pale brown with sparse spottings and blotches of black and umber.

Common Noddy   *Anous stolidus*

Gull-like in shape and size, but with tail longer and wedge-shaped, and entire upper surfaces dark brownish-grey except for the white forehead, and crown which shade back to grey on the nape of the neck. There are small white markings bordering the eye above and below, and the almost black wings, where folded, reach back to the tip of the equally dark tail.

The Common Noddy is an inhabitant of tropical seas, seen along Australian shores as far south as the Capricorn Group of islands off the Queensland coast, and the Abrolhos Islands of Western Australia. It spends all its time at sea except when nesting, taking its food (fish and squid) from the surface or by diving into the waters. Large flocks sometimes form, but this species does not appear to be migratory, and probably does not disperse far from the breeding islands. Large colonies form in the breeding season, which may be both in spring and autumn in some years. Nests within the colonies may be on the ground or on low shrubs, and built of twigs, seaweed or any other material available on the island. The single egg is white with brownish and reddish markings.

Crested Tern   *Sterna bergii*

The Crested Tern may be seen at any point around the Australian coastline; its range of distribution includes the entire Indian Ocean and through the islands of south-eastern Asia to Japan and to the central Pacific.

This large graceful bird is grey above and white below, and has a forked tail. During the breeding season the larger feathers of the black crown form a shaggy crest.

Compared with gulls, the terns (of which there are twenty-two species, including the Noddies) are more graceful and slender, with the wings narrower and the tail often forked.

Terns however hover well above the sea, then plunge-dive to take fish from below the surface, or in shallow swoop to take food from the surface.

The Crested Tern's nest is merely a slight depression in the ground; one egg is laid, being grey or brown with lines, smudges and blotches, mostly of black.

Pacific Gull   *Larus pacificus*

In general, shape resembles the smaller (length thirty-eight centimetres compared with the Pacific Gull's sixty centimetres) common Silver Gull. This big gull's bill is much deeper and heavier-looking, its wings and back are black, there is a black sub-terminal band across the tail, the legs are yellow instead of red, and the bill is yellow except for the red tip.

The Pacific Gull frequents the coastal waters and offshore islands around Australia's southern shores from the mid New South Wales coast to Point Cloates on the Western Australian coast. It feeds both from the sea, by diving and floating on the surface, and also on the shores. Nesting is usually in small colonies on islands, a nest of grass being built on the ground on beaches or among rocks. The two or three eggs are greyish with spots of brown.

HAWKS AND EAGLES

FAMILY *ACCIPITRIDAE*

The Australian members of this family include not only the hawks and eagles but also the kites, goshawks, harriers, and one species of buzzard. In this family the bill is always hooked, and the nostrils located in an area of soft leathery skin known as the cere at the base of the upper bill. The strong feet have one toe directed backwards and three forward, all with very long sharp claws. The eyes are extremely highly developed, with binocular vision in the forward direction.

All have the wings powerful and relatively large, the exact shape depending upon the mode of hunting — all use flight in hunting, whether by stealth, pursuit, by dive or 'stoop', or by dropping upon terrestrial or aquatic prey. Some soar for long periods when searching for prey, and have long broad wings that are held straight out from the body with flight feathers spread in a 'fingered' fashion. Others such as goshawks which rely upon stealth and a sudden dash, have shorter rounded wings, while the hovering kites have pointed wings. Always the powerful talons are used in the kill, and the prey is later ripped apart by the hooked bill. Of the world total of 2178 species Australia has 17, and 5 of these are endemic.

Black-breasted Buzzard
Hamirostra melanosternon

In many ways the Black-breasted Buzzard is an unusual bird of prey. It occurs only in Australia, and it is the sole species in its genus, although it has some similarities to the true buzzards of the genus *Buteo*.

The Black-breasted Buzzard, which occurs throughout Australia in open woodland scrublands and grasslands with scattered trees, is unusual in that it has two separate colour phases, a light and a dark. The lighter birds are mainly rufous brown with black streaks on the breast, while the darker phase is black on head, back, wings and breast and brown elsewhere. Distinctive features of both are the short square tail, and large white patch on the underside of the wings, which are held in an upswept position when soaring. Unlike the kite, the buzzard feeds more upon small animals than carrion.

Black-shouldered Kite   *Elanus notatus*

At first glance this bird of prey could be mistaken for a gull, being pure white and only slightly larger — but here any similarity ends. The Black-shouldered Kite preys upon small ground animals, including mice, lizards and insects. Its skill in hovering is equalled only by the Kestrel. The Kite can hold a position absolutely stationary relative to the ground, compensating for every gust of wind by a pause, or by faster, stronger beats, and the constantly changing angle of its tail. All the while its head is motionless. It gazes intently into the grass, then suddenly, wings folded back, drops down to take the prey.

The Black-shouldered Kite is very pale grey on the upper parts, with conspicuous large black shoulder patches, and black markings on the underwings; the head is white, with black above and behind the large red eyes. Among the birds of prey it could be confused only with the Letterwing Kite, which is almost identical except for the more extensive black underwing markings which in flight from below, form the letter 'M'.

The Black-shouldered Kite inhabits open woodlands and tree-studded grasslands, and has benefited from clearing of more heavily wooded country for farming. It occurs almost throughout Australia, but is absent from Tasmania.

Brahminy Kite   *Haliastur indus*

Also known as the Red-backed Sea Eagle, this kite has a wide distribution extending from Australia to south-east Asia and India. Its scavenging way of life and kite-like appearance make the name Brahminy Kite more appropriate; it is known by this name in India and Ceylon.

The colour pattern of this species is attractive, the head, neck and breast being pure white, the back, abdomen, wings and tail bright reddish-brown, and the wing quills black. The Brahminy Kite inhabits the northern Australian coastline, from the mid-west coast of Western Australia to northern New South Wales, frequenting mangroves and other trees around estuaries and the lower reaches of rivers, where it feeds mainly upon carrion and fish. The nest is a large platform of sticks in a tall tree overlooking sea or swamp

Collared Sparrowhawk
Accipiter cirrhocephalus

One of Australia's smallest birds of prey, the Collared Sparrowhawk has dark grey-brown upper parts with a rusty-brown collar around the back of the neck; the undersurfaces of body and shoulders are finely barred with cinnamon brown and white, while the undersurfaces of tail and wings are more boldly barred brown and pale grey. The eyes and legs are bright yellow.

The Collared Sparrowhawk hunts by sudden rush from some vantage point in a tree; or by dashing through the foliage to panic its prey of small birds and other animals. It is a bird of forests, woodlands and, through the interior, the timbered margins of watercourses.

A nest of sticks lined with green leaves is constructed in a high fork of a tree; three of four greenish-white or bluish-white eggs are laid.

Crested Hawk   *Aviceda subcristata*

The Crested Hawk is unusual among Australian birds of prey not only in its appearance, but also

in its behaviour, for it hunts insects among the foliage of trees, snatching cicadas and the like from the leaves.

Although its length of thirty-five to forty centimetres is not great, it gives the impression of being a larger bird than some other insectivorous hawks of similar length (such as the Black-shouldered Kite). Its long and broad wings give it a slow, buoyant flight, the value of which is appreciated when it is seen fluttering slowly through the treetops in search of its insect prey. Even the nesting material of green leaf twigs is gathered in flight, in the tree-tops.

This rather uncommon species occurs in Indonesia, New Guinea and the Solomon Islands; within Australia it inhabits forests and woodlands of the northern and eastern coasts.

Spotted Harrier   *Circus assimilis*

Harriers, of which Australia has two species (Spotted and Swamp Harriers) fly low across open plains, reedbeds and dry scrublands, often just two or three metres above the ground. Their flight is a lazy glide with an occasional leisurely flap of their long broad wings; they do not hover.

The prey of the Spotted Harrier consists of small birds and animals up to the size of a rabbit.

Crested Hawk   *Aviceda subcristata*
A Crested Hawk settles warily onto its nest fifteen metres above the ground in a slender eucalypt, watching with eyes of incredible bright yellow the camera lens which protrudes from the canvas 'hide'. This is one of Australia's most beautiful birds of prey, with its boldly barred breast and abdomen, bright cinnamon-brown plumage under the shoulders and tail, delicate blue-grey head and upper breast and, unique among Australian hawks, a conspicuous crest at the back of the head.

Tawny-crowned Honeyeater *Phylidonyris melanops*
On a scarlet *Banksia coccinea* flower a Tawny-crowned Honeyeater lifts his wings to fly, revealing the delicate pinkish-cinnamon tints normally hidden from view by his folded wings. These restless and rather elusive birds (they are often more easily heard than seen) are nomadic in some districts, wandering wherever the wildflowers are most abundant.

Right Above:

White-cheeked Honeyeater *Phylidonyris nigra*
A White-cheeked Honeyeater, wings half opened to preserve its balance, thrusts its long bill energetically into a cluster of *Beaufortia* flowers. This honeyeater closely resembles the more common White-eyed or New Holland Honeyeater, but can readily be identified by its much larger white cheek patches, and its brown eyes.

Right Below:

Western Spinebill *Acanthorhynchus superciliosus*
Clinging to the stem of a One-sided Bottlebrush (*Calothamnus*) a Western Spinebill probes among the crimson brushes of the flowers. It is in this species that the long curved honeyeater bill has reached its greatest development; Spinebills are well adapted to feed at the many deep narrow tubular and brush-like flowers of the Australian bushland.

Probably the bird's sudden appearance so close overhead panics many small creatures which would not otherwise reveal themselves.

The Spotted Harrier is one of Australia's most attractively plumaged birds of prey, with russet on the head, and most of the underparts brown boldly spotted with white. The eyes and the long legs are yellow. It occurs throughout Australia in suitable habitats.

Wedge-tailed Eagle   *Aquila audax*

Australia's largest bird of prey, the Wedge-tailed Eagle is found throughout this continent and occurs also in New Guinea. Wingspan averages two to two and a half metres, but measurements up to almost three metres have been recorded. The adult Wedge-tail is wholly black except for the rufous, black-streaked nape; immature birds are mostly dark brown with wing quills and tail black. The legs are feathered right to the powerful, whitish, massively clawed toes. Most distinctive is the long, wedge-shaped tail.

This eagle inhabits a great variety of country, from heavily forested coastal ranges to semi-desert inland plains. It is most often seen soaring on outstretched wings, usually at great height. But once in a while, beside some inland roadside, one of these magnificent great birds may occasionally be disturbed where it is feeding at a carcass. Lifting ponderously into the air it affords a closer look, a glimpse of the heavy hooked bill and huge powerful talons.

The Wedge-tailed Eagle builds an enormous nest of sticks in a tall tree, often on a hillside, or at the head of a gully, where it commands a sweeping view across the countryside, and can detect any intruder from far off. The one or two eggs are whitish, marked with lines and blotches of reddish-brown, and measure about seventy-three by fifty-eight millimetres.

Azure Kingfisher   *Ceyx azureus*
Immediately before diving into the entrance of its nest tunnel in a river bank the Azure Kingfisher uses its entire undersurfaces to catch the air and reduce its speed. This small kingfisher is only eighteen centimetres in length, and of this, the bill accounts for five centimetres. The kingfisher's prey is captured in a headlong dive into the shallow waters of a creek, river or swamp.

Whistling Kite   *Haliastur sphenurus*

Formerly known as the Whistling Eagle, this bird is obviously kite-like in its appearance and habits. It is more a scavenger than a hunter in the manner of the true eagles; its flight is slow, with much gliding and wheeling in circles, a lazy-looking flight on long upward-bowed wings. Its legs are bare like other kites, rather than feathered like the eagles.

The Whistling Kite is distributed throughout Australia, but is rare in Tasmania, preferring a more open habitat of dry forests and the open inland savannah-woodlands and semi-arid scrublands, where its prey consists of carrion, very small or weak animals and large insects.

The stick nest, constructed in a high fork of a tree, may be used for many years, and becomes very bulky with the new material added.

White-breasted Sea Eagle
Haliaeetus leucogaster

Sea-eagles are distinguishable from the true eagles by their bare instead of feathered legs. With a wingspan of around two metres these birds are only slightly smaller than the Wedge-tail, but easily distinguishable by their colouring. Mature birds have the head, neck and entire underparts, except for wing quills, pure white. The back, wings and tail are grey, and the wedge-shaped tail is broadly tipped white. Young Sea Eagles, being brown streaked with light and darker brown, could possibly be confused with the immature Wedge-tailed Eagle, but their white tail is distinctive.

White-breasted Sea Eagles inhabit the coasts, estuaries and lower reaches of rivers around Australia and Tasmania, and extend northwards through New Guinea to south-east Asia. The nest is a huge pile of sticks in a tree, on a cliff, or often on island or offshore rocky spire.

HONEYEATERS

FAMILY *MELIPHAGIDAE*

The honeyeaters are small to medium-sized nectar and insect-eating birds of the south-western Pacific region, occurring in Australia, New Guinea, New Zealand and as far eastwards as Hawaii. The one hundred and seventy species typically have a slender, usually long, down-curved bill, and a very long extensile brush-like tip. The sides of the tongue curl inwards to form a tube through which nectar and tiny nectar-drinking insects can be sucked up. Honeyeaters are abundant among Australian birds, and occur in almost every habitat; their success may be attributed to the abundance of flowering trees and other plants, while they themselves have undoubtedly contributed to the evolution of many of the unique wildflowers, some of which are entirely dependant upon the honeyeaters for their pollination. The honeyeaters family contains birds of very diverse character, as different in size and appearance as, for example, friarbirds and spinebills, so it is thought that the family may be a composite of birds of different origins which have all exploited the abundance of nectar as a food source. Australia has seventy species, of which fifty-four are endemic.

Helmeted Friarbird   *Philemon yorki*

These large honeyeaters (length 30 centimetres) have a most grotesque appearance. Their body plumage is an undistingished dull brown, white on the breast. But the head has an almost vulture-like appearance.

The sides of the face are bare of feathers; instead the skin is very dark, almost black, making the red eye all the more conspicuous. The bill is black, and a high crest-like ridge extends from the bill onto the forehead. There are in Australia altogether five species of friarbirds, which although similar to other honeyeaters in nectar and insect-seeking habits, are in appearance very different. The Helmeted Friarbird inhabits coastal north-eastern and northern Australia, where its loud harsh calls make it a conspicuous bird of the swampy heathlands.

New Holland Honeyeater
Meliornis novaehollandiae

This boldly plumaged bird is known by a variety of common names. 'New Holland' is derived from its scientific name, which translates as 'honey bird of New Holland', while the alternative 'Yellow-winged Honeyeater' and 'White-bearded Honeyeater' are purely descriptive. It closely resembles the following species, the White-checked Honeyeater.

The New Holland Honeyeater is a common bird, inhabiting heathlands, open forests and suburban gardens. It is restless, active and noisy, so that its presence in any numbers, as when banksias, eucalypts or other trees are flowering, can hardly be overlooked. The calls are loud, short and shrill. If a hawk or other predator is sighted the alarm call of one bird is immediately echoed by every other of these birds in the vicinity; the silent glide of a goshawk through the treetops is preceded first by an anxious ripple of warning sound. Every small bird dives for cover, and for a few minutes the bush seems deserted.

Tawny-crowned Honeyeater
Phylidonyris melanops

Principally a bird of scrubby heathlands, sandplains and mallee thickets, this honeyeater feeds upon the nectar of small wildflowers which are abundant in such country, and also captures insects at the flowers. Although its clear notes can be heard from a considerable distance, their ventriloqual quality and the bird's habit of keeping among the vegetation can make this honeyeater difficult to observe, except when one occasionally sits conspicuously on top of a bush or tree to call. During the breeding season the Tawny-crowned Honeyeater's display flights attract attention as it rises vertically to heights between 10 and 15 metres, then drops downwards while calling loudly. The nest is hidden in a small bush near the ground.

This honeyeater, an endemic species, occurs in south-western and south-eastern Australia.

Western Spinebill
Acanthorhynchus superciliosus

One of the most colourful of small Australian honeyeaters, the male has a broad bright cinnamon throat which contrasts with white and black crescent-shaped bands across the breast. The bright red eye is set in a black band.

It is in this bird that the curved honeyeater bill reaches its greatest development — this bird's beak (and that of the similar Eastern Spinebill) is extremely long and slender, perfectly suited to probing for nectar in the many small tubular-flowered plants of the bushland undergrowth, or pushing into the stiff brush-textured flowers of such trees and shrubs as *Banksia* and *Dryandra*.

The Western Spinebill is confined to forests, woodlands and heathlands of the south-western corner of the continent, where it is a common species.

White-cheeked Honeyeater
Phylidonyris nigra

Two confusingly similar honeyeaters of the Australian bush are the New Holland or White-eyed Honeyeater, and the White-cheeked Honeyeater. Both are dominantly black, boldly streaked with white especially on the breast; both have conspicuous bright yellow wing patches and yellow on the tail, and both have white markings about the head. The White-cheeked Honeyeater, however, has very large white cheek tufts, and the eye, so conspicuously white on the other bird, is brown.

The White-cheeked Honeyeater, the less common of the two, generally inhabits open heathlands, where it nests in the low shrubs.

IBIS AND SPOONBILLS

FAMILY *THRESKIORNITHIDAE*

Ibises are unmistakable with their very long thin downcurved bills, which are used for probing in muddy ground, swampy vegetation and shallow water. Plumages may be entirely dark, pied, or wholly black. In flight their long necks are carried fully extended.

Of the world total of twenty-six species of Ibis, three occur in Australia, and one, the Straw-necked, is unique to this continent (in actual fact a few individuals wander down Torres Strait to New Guinea).

Although of similar size, spoonbills have a bill that sets them apart from the ibises, perfectly straight and with the unique widened flat tip which is used to sift their food from the water. They are mostly white plumaged. Of the six species two occur in Australia and one, the Yellow-billed, is endemic.

Royal Spoonbill   Platalea regia

Ranging across most of Australia except the desert regions, the Royal Spoonbill is a large (seventy-five centimetres tall) white-plumaged bird with black bill and legs. It is generally seen feeding in shallow water using sideways swing-

ing and probing motions of the long spoon-tipped bill, searching for molluscs and other small forms of aquatic life. During the breeding season long white plumes develop at the nape of the neck. The nest may be built in a tree or upon trampled-down reeds, and three or four eggs, chalky-white with brownish markings, are laid.

Yellow-billed Spoonbill   Platalea flavipes

In general appearance resembling the preceding species, but with beak and legs yellow, and standing slightly taller. In the breeding season long white nuptial plumes appear on the breast. The bare skin of the face is blue with a pink tinge, but immature birds have face and bill pinkish yellow. Yellow-billed Spoonbills nest in colonies, building platforms of sticks in paperbark trees, bushes or reeds. This species is widespread in eastern Australia, but does not occur in Tasmania. In Western Australia its dis-

The nest of the White Ibis is constructed on trampled-down reeds, lignum bushes or other swampland vegetation; usually many birds nest together to form large colonies.

tribution extends southwards to the vicinity of Perth.

Straw-necked Ibis *Threskiornis spinicollis*

More strikingly coloured than the White Ibis, this species has its upper parts black, with bronze, purplish and green metallic highlights across the surface when seen close at hand in bright light. The back of the neck, the throat and the under parts are white, and there is a black band around the upper breast. The most remarkable feature of its plumage is the large tuft of yellow straw-like feathers that hang down the breast from the base of the neck.

The Straw-necked is the most widespread of the three ibis species in Australia; it is found throughout the mainland but is only a visitor to Tasmania. Like the White Ibis and Glossy Ibis, this species feeds in shallow water or wet grasslands, probing the bill into the water, mud or grass for insects and other small creatures.

White Ibis *Threskiornis molucca*

This is a species which Australia shares with New Guinea and Indonesia. Within Australia it is most common along the tropical northern and north-eastern coasts, but may be seen at favourable times in almost any part of Australia, including Tasmania. It is uncommon in the western half of the continent, south of the Kimberleys, only occasional individuals being seen usually in company with the Straw-necked Ibis.

This large bird is entirely white, with the exception of the wing quills which are black. The beak and the area of naked skin on head and neck are black, and the legs are pink on the upper and brown on the lower part.

In some areas where birds are more common flocks may be seen hunting for insects and small aquatic creatures on swampy grasslands.

JACANAS

FAMILY *JACANIDAE*

The lily-trotters, jacanas or lotus-birds are members of a very distinctive family of slender, long-necked birds which are usually regarded as being a specialized branch of the wader group. Their chief distinction is their toes which are exceptionally long to distribute their weight over the rafts of floating lilies and buoyant water plants of inland swamps.

Peculiarities of most members of the family include a metacarpal spur, like that on the bend of the wing of the plovers (which are probably the jacanas' closest relatives) and, in some species at least, reversal of male and female roles, the female displaying to the male in courtship, and the male doing most of the incuba-

tion and care of the young. There are altogether seven species in the family, of which one occurs in Australia and extends northwards to New Guinea.

Lotus-bird *Irediparra gallinacea*

This bird, also known as the Jacana, is unique among Australian birds in its appearance, and is quite unmistakable. Its domain in the swamps and lagoons is the floating layer of lily leaves, upon which it is able to walk, exceptionally long toes spreading its weight across a larger surface area of the fragile leaves. Although the total length of the bird is twenty-three centimetres, the legs are eleven centimetres and the longest toes, the cental front and the hind toes, are 7.5 centimetres long, giving each foot a fifteen centimetre span, sufficient to distribute the bird's weight evenly across the plate-like floating leaf of a waterlily.

The Lotus-bird has from its bill up onto its forehead a large fleshy orange comb; the face, front of the neck and upper breast are white blending into orange at the margins, the crown of the head, back of the neck and the breast are black, the belly and under-tail coverts are white, and the back and wings are brown.

This bird is found on calm inland waters where there is abundant floating vegetation; within Australia its range of distribution covers the northern and eastern coasts as far south as the Sydney region. Its nest is constructed of floating seeds and other vegetation, or consists of a large floating leaf. The three or four eggs, which float if tipped from the nest, are yellowish or pale brown patterned with black lines.

LARKS

FAMILY *ALAUDIDAE*

The lark family contains seventy-five species renowned for their song; their comparatively dull plumages make them visually insignificant. The majority of species occur in Africa, but representatives range across a large part of the world; only two species occur in Australia. The larks are birds of open areas, the deserts, beaches, grasslands and ploughed paddocks. Larks habitually sing on the wing, often while soaring. Anatomically they differ from other passerine birds in having the rear edges of the legs rounded instead of sharply ridged. There are usually crests or tufts on the head, long pointed wings, a conical bill, and the plumages are usually in browns and greys. Except for a few species that perch on bushes or posts, larks normally land only on the ground, where they spend most of their time searching for small insects.

Bushlark *Mirafra javanica*

This terrestrial bird may attract attention to itself with its display flights (which are rather like the flights of the European Skylark) when it rises steeply into the air, and at the peak of its flight remains poised on rapidly vibrating wings and singing loudly, before dropping back to the ground. It often flutters briefly above the grass before dropping down. The Bushlark runs swiftly; it does not have the habit of the rather similar Pipit of bobbing its tail up and down. The distribution of this species covers northern and eastern Australia, being absent from the region south of Shark Bay in Western Australia and west of Eyre Peninsula; it is also absent from Arnhem Land, Cape York, heavily forested parts of south-eastern Australia, and Tasmania.

The Bushlark in general is dark above, light below, with a distinct pale eyebrow line, and white edges to the tail. The breast is speckled with black, and there is a reddish patch on the wing. Compared with the Pipit it has a shorter tail giving a more squat appearance, and is more brownish in colour.

The Bushlark's nest is a hollow scraped in the ground and lined with grass. It may be an open cup or the grass may be pulled over to form a canopy. The three or four eggs are pale greyish freckled with dark grey or grey-brown.

KINGFISHERS

FAMILY *ALCEDINIDAE*

Most of the eighty-seven species of kingfishers are tropical birds, and their stronghold the islands of the south Pacific. They have short compact bodies and tight plumage. Their flattish feet have the middle and outer toes joined for much of their length, and they hop rather than walk. Tails are very short and bills very long and straight. The kingfisher family is divided into two main groups, the wood kingfishers (*Daceloninae*) and the true kingfishers (*Alcedininae*) The wood kingfishers have their headquarters in the Malayan-Australian region, and do not usually hunt fish, but prey on small land animals, especially frogs, small lizards, centipedes and other large invertebrates; the larger species such as the kookaburras include quite large snakes in their diet. The wood kingfishers have the largest and most unusual species; their bills are rather flattened, and with a hook at the tip. Australian wood kingfishers include the two kookaburras, the Forest Kingfisher, the Red-backed, Sacred, Mangrove, Yellow-billed and White-tailed Kingfishers. The second group contains the true kingfishers which have the bill straight and sharp-pointed, and have tight oily plumage; they are tied to rivers and streams, and hunt small fish and aquatic crustacea by diving into the water. In Australia this group includes only the Azure and Little Kingfishers. Altogether Australia has ten species of kingfishers, of which two are unique to this continent.

Azure Kingfisher   *Ceyx azureus*

Rarely if ever venturing away from the immediate vicinity of its inland waters habitat (creeks, rivers, estuaries and mangrove swamps), the Azure Kingfisher is most often to be seen on a perch overhanging the water, motionless but for a regular jerking of the stumpy tail and an occasional bobbing of the head. Should a small fish or crustacean be seen it will dive vertically, hitting the water hard and fast, beak first, and generally emerge a few seconds

later carrying its prey as evidence of its hunting skill.

The Azure Kingfisher within Australia inhabits the northern, eastern and south-eastern coastal regions, and extends far inland along major river systems. The species extends northwards to New Guinea and other islands.

LORIKEETS, COCKATOOS AND PARROTS

FAMILY *PSITTACIDAE*

These birds make up one of the best known of all bird families, which has spread to all of the warmer regions of the world. All species have the skull relatively large and rather hawk-like, with the nostrils set in a fleshy cere at the base of the bill, unusual in that the upper as well as the lower halves are moveable. A transverse 'hinge' at the base of the skull enables an up and down movement of the upper mandible, and makes possible a leverage between the upper and lower greater than with any other birds. Parrots and cockatoos are able to crack rock-like seeds and hack away hard wood, whether of nesting hollows or cage frames. Their feet are unusual, with two toes forward and two backwards; they are skilled climbers, and have almost human ability in use of the foot as a hand, particularly while eating. They demonstrate a remarkable ability to learn rapidly, and are famed for their powers of mimicry. Parrots are generally divided into two main groups, the parrots and cockatoos with blunt tongues, which feed on seeds and fruits, and those with brush-tipped tongues, the lorikeets, which feed on nectar and pollen. Some feed on the ground, others only in the treetops some take to the air infrequently, but most are strong fliers. Of the world total of 515 species, fifty-five occur in Australia, and no less than forty-eight of these are endemic.

Fig Parrot   *Opopsitta diopthalma*

Widespread and common in New Guinea and adjacent islands, the tiny Fig Parrot within Australia inhabits the rainforests and adjacent eucalypt forests of the north east, coming as far south as the Macleay River in New South Wales. The Fig Parrot is not only tiny, but departs from the typical long slender parrot shape by having a very short tail.

The upper plumage is dark green, the underparts paler green, the flanks yellow, the inner wing tipped with red and the outer wing with black and edged with blue. Three Australian races have been described, according to varying patterns of red, violet and blue about the forehead, eyes and face. Fig Parrots in small flocks fly and feed high in the forest canopy, where they are generally not easily observed. The nest is a small hole in a high branch, where two white eggs are laid.

Great Palm Cockatoo
Probosciger aterrimus

The Great Palm Cockatoo is a big black bird with a massive bill, and a head that appears all the

larger for its very long straggly crest. The only colour is the patch of bare red skin below the eye.

This is a New Guinea bird, which has established itself in Australia only at the tip of Cape York, where it frequents forests and adjacent woodlands, keeping mainly to the foliage canopy of tall trees, but descending to feed on pandanus palm seeds and other fruits.

A deep hollow, usually at a considerable height, is the nest site, where a single egg is laid.

Ground Parrot   *Pezoporus wallicus*

This unusual parrot is terrestrial in habits, rarely flying except when disturbed. It then flies swiftly and erratically, usually only for a short distance, before dropping to the ground again. It inhabits the swampy margins of coastal heathlands, and usually occurs in small colonies.

Crested Pigeon   *Ocyphaps lophotes*
As this Crested Pigeon settles onto its untidy nest it fans out its long tail; clearly visible is the iridescent green wing speculum, effectively set against the softly pinkish-grey plumage and the black-barred shoulders.

Western Silvereye   *Zosterops gouldi*
A Western Silvereye, one of the commonest of small birds, thrusts a whole (and probably living) spider down the throat of its young. The birds of this family, the Zosteropidae, of which there are altogether about 80 species (forty occurring in Australia) all have a distinct white eye ring.

The Ground Parrot is cryptically coloured, being principally grass-green, blotched and barred with yellow and black. There is a narrow red bar across the forehead, and a yellow bar across the wing.

These birds are not often seen, being active at dawn, dusk and during the night when they may undertake nomadic flights to other areas of suitable habitat.

The nest is a hollow scratched into the ground and lined with leaves, twigs and grass, and partly hooded over with stalks of ferns. This parrot, having suffered extensive loss of habitat, and being very vulnerable to introduced predators, has become rare.

Rainbow Lorikeet
Trichoglossus haematodus

The gaudily coloured Rainbow Lorikeet, also known as the Blue-bellied Lorikeet, Rainbow Lory and Blue Mountain Parrot, is a common species in coastal eastern and north-eastern Australia. The head is dark blue-violet, the collar around the neck is greenish-yellow, the breast yellow streaked with red, the belly very dark blue, and the undertail green and yellow. The wings and upperparts are a rich vivid green, and there is bright red and yellow hidden under the wings (but conspicuous in flight). The bill is reddish-orange and the eyes orange.

Rainbow Lorikeets are nomadic, wandering extensively in search of flowering trees, where they find pollen, nectar, fruits, seeds, berries, and also take insects from flowers and foliage. In the northern parts of their range of distribution they are present throughout the year, but in the south they are nomadic, being at one time absent, then suddenly returning in large numbers. The nest is a hollow in a tree, where two or three white eggs are laid.

Black Swan *Cygnus atratus*
Constructed on the waters of a shallow lake, where the massed reeds rising just above the surface give the impression of a lush green field rather than of water, the Black Swan's nest floats in a small clear space where the bird has gathered in the vegetation needed for its construction. This swan has just hauled itself up from the water, and is about to settle down onto the eggs.

Red-winged Wren *Malurus elegans*
A male Red-winged Wren throws his wings up to fly from his perch on an old banksia seed cone lying in the forest undergrowth; he carries a moth to be fed to young in a nest a few metres away. The chestnut shoulder patches which give this species its common name are at this moment hidden by his uplifted wings.

Red-tailed Tropicbird *Phaethon rubricauda*
A red-tailed Tropicbird broods its eggs on a ledge of the coastal cliffs at Sugarloaf Rock, near Cape Naturaliste, south-western Australia, the only mainland nesting place of these birds in Australia; offshore, it nests at Raine Island, on the north-eastern coast. Although this tropicbird is 45 centimetres in length, it has the two central tail feathers extended to form slender 'streamers' up to 30 centimetres in length.

Red-capped Parrot
Purpureicephalus spurius

Found only in the eucalypt forests of the south-western corner of the Australian continent, the Red-capped Parrot is the sole representative of its genus, and may be a relict form of a much more widespread group which has become extinct elsewhere. It is quite different in plumage pattern to any other Australian parrot, and with deep violet, bright yellow, red, blue and green is one of the most colourful of parrots.

The Red-capped Parrot is also of interest in its close association with a species of eucalypt, the marri, which is common within the parrot's range of distribution. Using an unusually long-pointed upper bill (presumably evolved specifically for this purpose) the Red-cap is able to probe out the seeds from the very hard wooden capsules (gumnuts) of this tree. In this respect it has a distinct advantage over other species of parrots which must spend a great deal of time chewing through the very dense wood of the seed capsule if they are to make use of this very abundant source of food.

LYREBIRDS

FAMILY *MENURIDAE*

Of all the vast array of perching or song birds, which make up two thirds of the world's birds, the largest and one of the most remarkable groups is that comprising Australia's two species of lyrebirds. In its shape the lyrebird resembles the gallinaceous birds, especially the megapodes, but it has a primitive syrinx like the 'voice boxes' of the song birds. It has also an elongated breast-bone which is quite distinct from anything else known but has affinities to the sterna of perching birds; however it differs from them all in having only rudimentary clavicles. Its head is small for its body, the legs long and powerful, with large fowl-like feet. These characteristics, and the unique tail, resulted in the placing of lyrebirds with the perching birds, but in a distinct family of their own. The tail of the Superb Lyrebird, the more highly developed of the two species, is enormously long, and consists of the unusual number of sixteen retrices. The outermost tail feathers, some fifty centimetres in length, are whitish with prominent brown 'V' shaped notches. The lyrebird family is confined to Australia.

Albert Lyrebird *Menura alberti*

Closely resembling the Superb Lyrebird, but more rufous toned on the back, and tinged with brown on the underparts, particularly the undertail coverts which are chestnut. The outer lyre-shaped tail feathers are absent.

The Albert Lyrebird inhabits mountainous forest country of south-eastern Queensland, mainly to the north of the range of the Superb Lyrebird. It is quite common within the rather restricted areas of suitable habitat.

The male Albert Lyrebird does not construct a mound or clearing, but displays from a log or

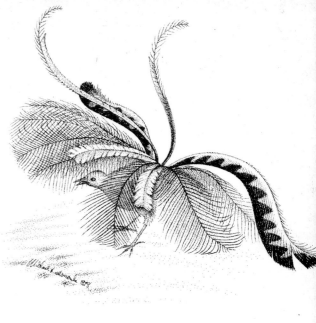

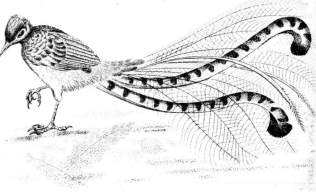

stump. As with the Superb Lyrebird, the nest is a roughly-constructed, domed structure of sticks, bark and grass, usually placed on a rock ledge or among logs, and only rarely built on the ground.

Superb Lyrebird *Menura superba*

Famed for its long decorative tail, the Superb Lyrebird is becoming equally well known for its

displays, and its vocal mimicry of the calls of birds, and various other forest sounds.

This large bird (length eighty to ninety-five centimetres, including tail forty-five to fifty centimetres) is principally brownish above and grey-brown on underparts, and normally is not particularly striking in appearance when the long tail trails behind. But when the male displays at his mound in a forest clearing, raising his tail at first vertically and then arching it forward above his body, all the fine filaments hang like a shimmering silvery curtain beneath which he is almost lost from sight. At the same time he sings with a powerful clear voice, some notes musical and others harsh, some purely his own and others in imitation of other bird songs.

Any female lyrebird in the vicinity is likely to be attracted to the mound, and mating may occur. Nest building and care of the young are left entirely to the female. Superb Lyrebirds occur in coastal mountainous south-eastern Australia from Victoria to southern Queensland.

MOUND BUILDERS

FAMILY *Megapodiidae*

The megapodes are the only known birds that make use of a source of heat other than the warmth of the sitting bird's body for the incubation of their eggs. Their young are hatched beneath a massive mound of warm earth and forest litter. There are several sources of heat. Most commonly utilized is the warmth generated by rotting vegetation, but some species use the heat of the sun upon a mound of sand to supplement the action of fermentation, and some even use volcanic heat, burying their eggs in the ground warmed by subterranean volcanic activity. They are sometimes known as 'thermometer birds' because when regulating the mound's temperature (by adding vegetation or hot sand, or opening the mound for cooling) the bird tests by contact with the tongue or skin on the underside of the bill; they are amazingly accurate in maintaining the optimum incubation temperature.

The chicks hatch buried deep in the mound, struggle to the surface, and run off into the forest, quite able to look after themselves; parents and chicks show no interest or recognition of each other. There are thirteen species extending north to Malaysia. Of these, three occur within Australia, and two are endemic.

Brush Turkey   *Alectura lathami*

A large black bird which has the bare skin of head and neck coloured bright red, becoming yellow towards the base of the neck and on the wattles.

The Brush Turkey is one of the megapodes, which build up a mound of rotting vegetation. This generates heat, and serves as an incubator. The birds spend a great deal of time attending to the mound in order to maintain a constant optimum temperature. The chicks emerge fully fledged and capable of looking after themselves, and are not assisted or fed by the parents.

Brush Turkeys are inhabitants of forest, from coastal rainforests to drier brigalow and mulga. The genus, of which this is the only species, is restricted to Australia, where it occurs in the east and north-eastern parts.

Mallee Fowl   *Leipoa ocellata*

Most of the megapode family — those fowl-like birds which hatch their eggs in huge incubator mounds of decaying vegetation — are birds of tropical rainforests, where abundant leaf litter and a warm moist climate are ideal for this type of nest. But Australia's unique Mallee Fowl has modified the megapode incubation for the dry eucalypt woodlands and mallee eucalypt scrublands. In this environment the initial source of heat, when the mound is still moist from winter rains, is from the decaying vegetation. But as the hot dry summer approaches and the mound dries out, halting the fermentation process, the direct sunlight on the mound becomes the dominant source of heat. The daily routine of opening out the mound to let in the sunlight, and piling back on the hot sand later in the day requires the constant attention and a great deal of labour on the part of the male bird.

The Mallee Fowl, a bird somewhat larger than the ordinary domestic fowl, is boldly patterned in cryptic coloration of brown, grey, black and white. It occurs across most of southern Australia except very heavily forested and mountainous areas.

Scrub Fowl   *Megapodius freycinet*

About the size and general shape of an ordinary domestic fowl, dark brown above and grey beneath, and with its only adornment a crest at the back of the head, the Scrub Fowl appears to be a most undistinguished creature. But it is a member of the Megapode Family, one of the mound-builders, and thereby becomes of considerable interest through its use of fermenting vegetation to incubate its eggs in the manner of the Brush Turkey. The mound is built up of inorganic materials upon which is piled up a dense

layer of rotting leaf litter. In the moist warmth of the tropical rainforests and monsoon forests which are its home this method of incubation works well. Some mounds accumulate over a long period of time, may be up to four metres high and fifteen metres across. Within Australia, this species inhabits coastal Northern Territory and coastal north-eastern Queensland.

MONARCH AND ALLIED FLYCATCHERS

FAMILY   *MONARCHIDAE*

The birds of this family are sometimes included with the Muscicapid flycatchers, with which they have some features in common. All have wide bills and well-developed bristles fringing the mouth; most are rainforest species inhabiting tropical northern and north-eastern Australia. The monarch flycatchers (genus *Monarcha)* actually take their insects on foliage more often than in the air, but have typical flycatcher broad bristle-edged bills. They are distinctively and beautifully plumaged in grey, black and rufous but for one species, which is black-and-white only. Another group of flycatchers of this family, those of the genus *Myiagra,* resemble the Willie Wagtail and other fantails, with long tails which they often swing or vibrate. Plumages are bold black-and-white or wholly black; some have areas of brown. The two species of the genus *Arses* have distinctive broad white frill-like collars and fleshy eye-rims; they inhabit the north-eastern Queensland rainforests and find much of their insect food in a most un-flycatcher fashion, on the tree trunks. The other genus of this family, *Machaerirhynchus,* is a New Guinea group of which only one species extends its range south to reach Cape York. This is the unique Boat-billed Flycatcher; like others of the genus it has its bill wide and flat, unlike those of any other small birds. Twelve species of this family occur in Australia, and five are confined entirely to Australia.

Pearly Flycatcher   *Monarcha melanopsis*

Mainly pearly-grey, with forehead and throat black, forming a mask (which does not include the eye). The face, around the eye, is almost white, emphasising the black, and there is a very thin black ring around the eye. The breast is grey, the abdomen, under the shoulders and under tail-coverts rufous.

This inhabitant of forests and woodlands, despite the 'flycatcher' name, is relatively slow-moving and only rarely catches insects in flight, but occasionally takes insects from leaves while hovering. It generally feeds among the foliage and along the branches. It is distributed along the full length of Australia's eastern coast from Cape York to Victoria.

The Pearly Flycatcher nests rather high up, in dense foliage, usually blending the deep cup-shaped nest into a vertical fork. It is made of various rainforest mosses and fibres, and camouflaged with green moss on the outside. The two eggs are white spotted with red.

THE MUD-NEST BUILDERS

FAMILY *GRALLINIDAE*

In New Guinea and Australia occur four species of birds which all employ an unusual nest construction technique. The nests are massive deep bowl-shaped earthenware structures, built of mud or clay reinforced with grass, horsehair or other fibrous material, and softly lined with grass and fur. They are built on a horizontal limb, often over a creek or river. There are within this family three genera of birds all with similar nests of mud but themselves quite different in appearance. These are the Mudlark (genus *Grallina*), the White-winged Chough (*Corcorax*) and the Apostlebird (*Struthidea*). The latter two genera of birds have a distinctive behaviour in that they form social parties of five to twenty birds, and do everything together — nest building, rearing of young, and roosting. Three of the total four species occur in Australia, where all are endemic.

Mudlark *Grallina cyanoleuca*

A boldly patterned black and white bird which derives this common name from its use of mud as a nest-building material. The nest is thick-walled, deeply cup-shaped, and large for the size of the bird, measuring about fifteen centimetres in diameter, with walls two centimetres in thickness and an internal depth around ten centimetres. This heavy structure, of mud reinforced with grass and softly lined with grass and feathers, is rather conspicuously placed on a thick horizontal limb or horizontal fork, generally at a considerable height.

Male and female Mudlarks have a duet song, one giving a 'te-he' call and the other following on immediately with a skillfully synchronized 'pee-o-wit', so that the call seems to have been given without pause, by a single bird.

Mudlarks are common and widely distributed in open woodlands, farmlands, parks and gardens almost throughout Australia.

NIGHTJARS

FAMILY *CAPRIMULGIDAE*

The Nightjars are plumaged in protective colours, which match fallen leaves and sticks so well that the birds seem to disappear when on the ground. In colours they resemble to some extent the frogmouths and owlet nightjars, being mostly tones of browns and greys, spotted, blotched and barred with lighter tints and darker shades of the same colours, sometimes with patches of white on throat and neck, as well as on wings and tips of tail feathers. Being nocturnal hunters of insects, the eyes are very large, while the heads are wide and rather flattened. The bill is very short and wide; when opened it resembles a gaping purse. Many species hunt by catching insects in the air. By day nightjars sit on the ground with eyes closed and blend so perfectly with their surrounds that they are almost impossible to detect. Of the world's sixty-seven species, three occur in Australia, and one, the Spotted, is restricted to Australia as a breeding species.

Spotted Nightjar *Eurostopodus guttatus*

Found throughout Australia except for the most heavily timbered areas and Tasmania, the Spotted Nightjar is seldom seen owing to its nocturnal habits; occasionally one may be flushed from the ground.

The plumage is richly marbled with chestnut and grey over the dark brown; the under surface is chestnut finely barred with brown, and with a sharply defined white arrow-shaped throat patch. On each wing, towards the tip, is a large, more or less rounded white spot. The beak, although much smaller and more delicate looking than those of the frogmouths, has a wide gape well suited to catching insects at night.

The single egg is laid on the ground among fallen leaves or other debris, and its colours, usually greenish spotted and blotched with tones of brown make it very difficult to see.

ORIOLES

FAMILY *ORIOLIDAE*

The true orioles consist of twenty-eight species of colourful, medium-sized birds that occur from Europe, Asia and Africa to New Guinea and Australia; they are not closely related to the orioles of the Americas. The orioles of this family are purely arboreal birds; all have quite long wings and straight bills. Yellow, green and black are the prevailing colours of most species. There are two groups, the orioles (genus *Oriolus*) and the fig-birds (genus *Sphecotheres*). The fig-birds number four species, confined to Australia and New Guinea, and distinguished by a patch of brightly coloured bare skin on the face. The other twenty-four species of the family are all known as orioles. Two species of fig-birds and two orioles occur within Australia.

Southern Figbird *Sphecotheres vieilloti*

This colourful bird feeds largely upon the fruit of wild fig trees in the rainforests of eastern and north-eastern Australia. Other fruits are taken, and it is a common visitor to gardens of towns within its range of distribution.

The male figbird is glossy black on the head and face, with bare skin around the eye a bright orange-red, and very conspicuous surrounded by black. The back of the neck, throat and breast are grey, the back and abdomen greenish-yellow, fading to white at the under tail-coverts. The flight feathers are black with grey edgings, the tail black tipped with white on the outer feathers. The eyes are dark red. The female is considerably less colourful.

The figbird builds a rather flimsy open woven basket-like nest in a horizontal fork of a tree, often quite high. The three eggs are greenish with brown spots.

THE OSPREY

FAMILY *PANDIONIDAE*

The single species that makes up this family is a large hawk which occurs almost throughout the world in tropical and temperate regions. (There are five slightly differing geographical races). The Osprey feeds solely upon fish, and differs from other hawks in having the outer toe reversible, giving two toes (and claws) both front and back, a better grip of the slippery prey. Although chiefly a bird of ocean coasts it may also be seen inland along major rivers and on large lakes. Male and female are alike, with the lower parts, head and neck mainly white, and the back and wings chiefly dark brown. The chest is banded with dark spottings, and the head usually has an almost black crest and dark eye streak. The female, sixty centimetres in length, is slightly larger than the male.

Osprey *Pandion haliaetus*

The Osprey or Fish-hawk has an almost worldwide distribution, and occurs on most parts of the Australian coast. It lives exclusively on fish caught by diving feet-first from a considerable height, sometimes hitting the water at such speed to plunge beneath the surface. The feet have very long and sharp talons, and the toes have a spiny undersurface to give a secure grip of the slippery prey.

The nest is a huge pile of sticks, usually on a small rocky islet or on cliffs at the tip of a promontory. Occasionally a mangrove or large bare-limbed tree will be used. Additional sticks and seaweed are added to the nest each year until it may grow to a height and diameter of several metres.

OWLET NIGHTJARS

FAMILY *AEGOTHELIDAE*

The owlet nightjars constitute a distinct group related more to the nightjars and frogmouths than to the owls. Yet they look rather like very small owls, sitting on tree branches with the erect posture of an owl rather than the more horizontal posture of frogmouths and nightjars; they could not be mistaken for owls, however, because of their long barred tails. During the day they sleep in tree hollows, and may sometimes be seen sitting at the entrance. The owlet nightjar family is restricted to the Australian-New Guinea region, where eight species occur, but only one of these within Australia; there is only the one genus within the family.

Owlet Nightjar   *Aegotheles cristatus*

Found throughout Australia, this small nocturnal bird is rather like a miniature frogmouth, with a rather owl-like face. It is a rather plump-looking bird, yet has a long barred tail. The upper parts are ashy-grey, the head is mottled with black and there is a narrow dark collar on the hind neck. The tail and wings are barred with brown, the underparts are finely barred grey, white and brown. The beak is rather like that of the frogmouths, very short and broad, and with bristles at the base. Like the frogmouths, the Owlet Nightjar does much of its nocturnal hunting on the ground, as ground-dwelling insects such as weevils, millipedes and ants predominate in the diet. However some insects are also taken on the wing.

The Owlet Nightjar roosts during the day in a hollow in a limb or the trunk of a tree. Such a hollow serves also as the nest, where three or four white eggs are laid.

OYSTERCATCHERS

FAMILY *HAEMATOPODIDAE*

The six species of oystercatchers, of which two occur within Australia, are large and boldly coloured shore birds. All are predominantly black or black-and-white, with brightly coloured legs and bills. Their most distinctive feature is a long straight spearlike bill which is laterally compressed almost like a knife blade, with a tip like a chisel. Its main function is to spear clams, mussels and oysters, but it is also used in capturing and killing marine worms and crabs. The oystercatchers feed when receding tides uncover reefs, rocks and mudflats. Their technique in dealing with the well-protected shellfish is to drive the knife-blade bill between the two halves of the shell, still slightly open in the very shallow waters of the receding tide, killing the animal within and causing the shells to fall open.

Pied Oystercatcher
Haematopus ostralegus

Oystercatchers have long straight bills that terminate in chisel-like points, used for probing, for prising shellfish off rocks and opening them. The Pied Oystercatcher which may be seen on any part of the Australian coastline, is black except for a white breast, belly, rump, upper tail coverts, and a white wing bar which is very conspicuous on the flying bird. The bill, eyes and legs of this forty-five centimetre long bird are bright red.

In habits the Pied Oystercatcher differs from the Sooty in its preference for sandy beaches and mudflats rather than the rocky coasts; here it uses its long bill for probing for its food in the sand or mud. The nest, a shallow hollow in the sand, contains two dull white, brown-spotted eggs.

Sooty Oystercatcher
Haematopus fuliginosus

The Sooty Oystercatcher is entirely a dull sooty black, against which the red bill and eyes, and the pink legs are a bright colour contrast. Unlike Australia's one other species, the Pied Oystercatcher, it feeds mostly on rocky sections of the coasts, and although uncommon, occurs around all parts of the Australian shoreline.

The Sooty Oystercatcher does not build any nest, but lays its two brownish or olive, grey and brown spotted eggs among seaweed, rocks, or on the bare beach sand.

THE PAINTED SNIPE

FAMILY *ROSTRATULIDAE*

Painted Snipe are aberrant relatives of jacanas and waders; there are two species, which have a very wide distribution in tropical and subtropical regions around the world. They are superficially snipe-like, but differ greatly from true snipe (family *Scolopacidae)* in anatomy and behaviour. Structurally Painted Snipe differ in having hard inflexible bills and wings that permit only a slow, rail-like flight with legs dangling below. They skulk in marshy vegetation in preference to the open beaches. An anatomical peculiarity is that the trachea of the female makes several loops before reaching the lungs, an adaptation most strongly developed in the Australian population, where there are as many as four loops. Consequently the female's voice is deep and resonant, while that of the male, which has an ordinary straight windpipe, is but a faint chirp. The female is also the more colourful, and these differences are in keeping with the reversed sexual roles. The females take the initiative in courtship and defend the territory, while the male builds the nest, incubates the eggs and cares for the young. This system has the advantage that the female, which is polyandrous, can lay many sets of eggs in a season, each set cared for by a different male, and thus a higher reproductive rate is obtained for the species.

Painted Snipe   *Rostratula benghalensis*

A colourful terrestrial bird, twenty-five centimetres in length, with a straight, slender and quite long bill.

The female is more colourful. There is a conspicuous white ring around the eye and a white streak between the back of the neck and the black wings, extending forward to join the white breast. The wings are dark olive-green barred with black, and the belly is white. The male differs in having more greyish foreparts in place of the black of the female's head, neck and upper breast plumage.

The Painted Snipe has a wide distribution, occurring in most parts of Australia, but probably absent from the northern parts of the Northern Territory and Cape York. Its habitat is the marshy shallows of swamps, where it nests on the ground or in the base of a bush near the water's edge. The four eggs are white with fine black lines.

PELICANS

FAMILY *PELECANIDAE*

These very large birds are found in many countries and their general appearance is well known. There are altogether eight species, of which only one occurs in Australia. Their most distinctive feature is a very long, straight bill beneath which is suspended a pouch. The bare skin of this pouch is used also in regulating body temperature. The pelican feeds by reaching down into the water and scooping with expanded pouch. The heavy body is carried on short, stumpy legs and large feet, which have all four toes united with a web to serve as swimming paddles. The other species of the family have a wide distribution, largely in equatorial regions, in the Americas, from Africa to south-east Asia and New Guinea.

Pelican   *Pelecanus conspicillatus*

The Pelican is distributed throughout Australia wherever suitable areas of water occur, whether temporary or permanent pools, rivers or coastal estuaries. Pelicans gather together in flocks, which spread out somewhat while feeding — which is done as the bird floats on the water by dipping the long bill and neck below the surface.

Nesting occurs whenever conditions are suitable in the drier parts of the interior, but in the south follows a regular spring (eastern areas) or autumn (western) pattern. The nest is on the ground, often on a small island, and built up of sticks and other vegetation; two or three white eggs are laid.

PENGUINS

FAMILY *SPHENISCIDAE*

Penguins pursue their food of fish, cephalopods, and various crustacea almost exclusively underwater. As a result they show extensive adaptations which make them completely unlike other birds and unmistakable in appearance. Their body shape has become seal-like. Propulsion through the water is solely from the flippers, modified wings with which they 'fly' through the water; the webbed feet and small tail serve to steer. Because the feet are set far back on the body, penguins on land adopt an almost vertically upright stance to maintain balance. All penguins have white fronts and dark backs, so that, seen from below, underwater, they match the silvery surface above, or from above, blend with the dark blue-greys of the ocean. Penguins are a prominent element of the antarctic and subantarctic faunas, but are also widely distributed in southern temperate oceans. Most are uncommon non-breeding visitors to Australia; one, the smallest of the penguins, breeds in Australian waters.

Little Penguin   *Eudyptula minor*

This penguin, the smallest of the world's eighteen species, is the only one to breed in continental Australian waters. Like the other species, the Little Penguin takes its prey of fish, crustacea and cephalopods underwater. Swimming is done with the powerful flipper-like wings which move up and down as do the wings of other birds in aerial flight. In the denser medium the relatively small wings of the penguin give strong forward thrust; it is literally flying through the water.

This species, also known as the Fairy Penguin, comes ashore only at night, being seldom seen during the day, when it is either crouched over its eggs in a burrow or rock crevice, or feeding at sea.

The Little Penguin is an inhabitant of the coastal seas of south-eastern and south-western Australia. It breeds on a great many of the small offshore islets, building a nest of seaweed, sticks and other debris in a short burrow, under a rock, log or dense shrub.

PETRELS, SHEARWATERS, PRIONS

FAMILY   *PROCELLARIIDAE*

The many sea-birds of this family wander the oceans and come to land only to breed. Most are medium-sized birds with the exception of the Giant Petrel. They are completely adapted to life on the seas, being skilled in the use of wind in energy-saving wave-top soaring. Their bills are distinctive, having a deeply curved 'nail' on the upper surface, towards the front, followed by a shallow hollow or saddle, then a single conspicuous (internally divided) nostril tube. About thirty-six species occur on the seas around Australia. Some of these never come within sight of land except as windblown casualties of storms, but six species breed on Australian shores and offshore islands.

Fairy Prion   *Pachyptila turtur*

Five species of prions are occasional visitors to Australian shores from their home territories in sub-antarctic seas, and one species, the Fairy Prion, is a resident and breeds in Australian waters.

The Fairy Prion, like others of its genus, has long and pointed wings, giving it a rapid erratic flight low across the surface of the water, sometimes skimming the surface with bill tip submerged to gather plankton. It has a wingspan of about seventy-five centimetres, and is blue-grey on upper surfaces with a broad band below the eye; the underparts and a band above the eye are white.

Breeding occurs on islands in Bass Strait, and the birds wander around the coast to the southern parts of Western Australia and the coast of southern Queensland. The nest is a burrow or a cavity among rocks lined with leaves; one white egg is laid.

Gould Petrel   *Pterodroma leucoptera*

This small petrel of the southern seas breeds on Cabbage Tree Island, near Port Stephens on the New South Wales coast. This and two other petrels have a distinctive 'M' shape across their upper wing surfaces formed by a dark leading-edge of the outer wing and a dark diagonal across the inner wing towards the tail. Of these three (the others being the Cook Petrel and the Black-winged Petrel) the Gould Petrel can be distinguished by its black crown and nape. The upper parts are dark grey, and under surfaces mostly white.

This petrel visits land only for breeding. The one white egg is laid beneath protecting boulders, fallen palm fronds or other vegetation.

Great-winged Petrel
Pterodroma macroptera

Most of the petrels which roam the oceans around Australia do not come ashore, being non-breeding visitors from antarctic seas, but two species, the Great-winged Petrel and the Gould Petrel, nest on Australian coastal islands.

The Great-winged Petrel has a wingspan of about one metre, and a body length of forty centimetres; the wings are broad and the tail long. Petrels of this genus can be distinguished by their relatively short bills, on which the short nasal tubes are slightly up-tilted. The general overall colour is brownish-black, with a black patch in front of the eye, and sometimes a small area of white on forehead and chin. The bill and legs are black.

This petrel visits the land only when breeding; when at sea it feeds upon squid. It breeds on islands along the south coast of Western Australia between Cape Arid and Albany and in the Great Australin Bight. A single white egg is laid, usually in a short burrow, otherwise under rocks or dense plant cover.

Short-tailed Shearwater
Puffinus tenuirostris

Also known as 'mutton-birds', these shearwaters nest in great numbers on many islands along the coasts of New South Wales, Victoria, Tasmania and South Australia. The islands of Bass Strait are major breeding grounds, where this species has long been commercially exploited. Young birds are taken from their nests, salt-cured and used for food, important by-products being fat, oil and downy feathers. In the 1968 season, for example, 466,200 birds were 'harvested'.

One characteristic of shearwater behavior is their nesting in colonies which on some islands contain as many as two hundred thousand birds within an area of perhaps a hundred acres. The regular timetable followed by the shearwaters during the breeding season is remarkable. The birds invariably arrive at the nesting islands during the last week of September, prepare nest burrows, mate and lay their eggs. The timing is the same for all colonies from the Great Australian Bight to the New South Wales coast and to southern Tasmanian islands. The nest is a tunnel, in which the nestling, once it is two or three days old, is left unattended during the day, and fed by parents each night.

PHALAROPES

FAMILY   *PHALAROPODIDAE*

Phalaropes are a small group of waders that have left the shores for the oceans — they could almost be described as 'swimming plovers'. Yet in behaviour they retain many sandpiper-like habits, particularly in their flocking movements when masses of these birds rise in tight formation, wheeling in the air, then land in a mass all at the same moment on the water and float in a tight raft. Phalaropes are rather small, silent birds with long necks. For their life on the seas there are adaptations not found on the shoreline waders. The feet are rather like those of grebes, with flexible membranes extending from the sides of the toes to make them more efficient as paddles. The tarsi are laterally flattened to enable the legs to cut through the water with less resistance as they move back and forth while paddling. Air bubbles trapped in the thick plumage provide greater than normal buoyancy, enabling the bird to ride high on the water. These birds breed in arctic and sub-arctic regions and wander as far south as Australian seas; two of the total of three species occur on coasts here occasionally as a result of storms at sea.

Red-necked Phalarope   *Phalaropus lobatus*

Three species of phalarope may be seen on Australian shores or seas. These birds breed in arctic and sub-arctic regions, and afterwards disperse widely over the oceans. Of the Red-necked Phalarope, there are a few records of its being seen along Australia's south-eastern coast. It is a very attractive small bird especially in breeding plumage, when it has a contrasting colour pattern of grey-black crown, back, hind-neck and breast, bordered at the front and up to the eye with chestnut, conspicuous against the white throat. In Australia however it would be seen in non-breeding plumage, when it has dark grey upperparts, with a black band through the eye. The underparts are white, striped with black.

This and the Wilson's Phalarope, the only other species recorded on Australian shores, is a bird of the open ocean, the ocean beaches and tidal pools.

PHEASANTS AND QUAILS

FAMILY *PHASIANIDAE*

A very large family containing many pheasants, quails, partridges and others; but within Australia it is represented by only three species (out of a total of 165). A number of quail, pheasants and peafowl have been introduced and live wild in various places. These are heavy-bodied, ground-dwelling birds which live in dense (often very low) cover and use flight mainly as a means of escape or for nomadic travels. The quails of this family are known as the 'true quails' and have four toes (bustard quails of the following family have three). They have very compact rounded but powerful wings which propel them explosively into the air when any intruder approaches too closely to their hiding places in the grass. One species is unique to Australia.

PIGEONS AND DOVES

FAMILY *COLUMBIDAE*

The names 'pigeon' and 'dove' are used interchangeably, there being no scientific distinction between the two; 'pigeon' tends to be applied to the larger species, but there are many exceptions to this. Some of the features of this family include short and rather delicate bills with the nostrils located in a large soft fleshy cere at the base of the bill, and plumage colours which are predominantly pastel, but with areas of brighter iridescent hues. Although many species feed on the ground, the legs are very short. Those that inhabit arid inland areas feed on dry seeds, and must find water regularly, usually at sunrise and sunset; other species live in rainforests, where they obtain native fruits. The birds of this family are extremely adaptable, and have spread into many different habitats. Powerful flight is evident in most species, often with an almost quail-like explosive take-off accompanied by a loud clapping sound from the wings. The young, for the first week or so, are brooded by both parents, and fed a regurgitated substance known as 'pigeon's milk', given by the adult inserting its bill into the wider bill of the young. Of the world total of 289 species, twenty-three are native to Australia, and fourteen are found only in Australia.

Crested Pigeon   *Ocyphaps lophotes*

This bird, which is widely distributed across Australia from eastern to western coasts but avoids the more heavily forested northern, south-eastern and south-western regions, is an inhabitant of woodlands and open scrublands.Its most distinctive feature is a thin black head crest; otherwise it is a greyish, yet quite attractive bird.

The Crested Pigeon is a common species which forms small flocks when feeding on the ground, or when visiting waterholes to drink at sunrise and sunset. At other times of the year, when breeding, usually September to February, it is a solitary species. The nest, built in a bush or tree, is a rough flimsy platform of sticks upon which the two white eggs balance precariously.

Diamond Dove   *Geopelia cuneata*

This small dove (length 20 cm) inhabits the drier savannah woodlands, grasslands, mulga-scrub plains and semi-desert country across northern Australia, as far south as northern New South Wales.

The name 'Diamond Dove' comes from the sprinkling of white spots over the grey-brown wings and back. The crown of the head, nape, and breast are bluish-grey, and the tail is grey with the outer feathers tipped white. The iris and the naked skin around the eye are red, and the legs pink. The nest is a frail platform of twigs and dry grass in a bush or small tree, in which two white eggs are laid. Cycles of abundance followed by complete absence have been recorded, probably resulting from varied seasonal conditions.

PIPITS AND WAGTAILS

FAMILY *MOTACILLIDAE*

This family is made up of two main groups, the pipits, of which only one species occurs in Australia, and the wagtails. These wagtails are not related to the Willie Wagtail (which belongs to the flycatcher family); the three species recorded in Australia are all rare vagrants from the northern hemisphere. Both the pipits and the wagtails are mainly terrestrial birds of slim build, with long tails, long legs, and sharp-pointed bills. They are insectivorous, and inhabit open grasslands and paddocks. They indulge in a peculiar type of tail-wagging — the rear half of the body as well as the tail is bobbed up and down. These movements are made more or less continuously as the birds forage among the grass, and are performed increasingly if the birds become agitated, as when a predator is sighted. While the pipits have camouflage plumage of streaked browns and greys, the wagtails of this family are more colourful. The three species of wagtails which on rare occasions have been recorded in Australia (the Blue-headed, Yellow-headed and Grey Wagtails) have blue, bright yellow, olive green, black and white in their plumage patterns. Australia has only one resident species, the Pipit, of this worldwide family of 48 species.

Pipit   *Anthus novaeseelandiae*

The little Pipit (length 16 centimetres) is a terrestrial bird which runs very fast across the ground in short bursts. The flight is undulating, with many changes of direction, and of short duration, the bird soon dropping abruptly to the ground. It inhabits open plains, grasslands and paddocks, and its plumage blends well with the dry grass and earthy colours of the inland plains.

The upper parts are brown, with darker streaks, and the tail has white outer feathers. Underneath, the throat and abdomen are whites, and the breast streaked with light buff and brown. There are numerous variations of this colour pattern, which is to be expected of a bird that occurs throughout the continent.

A cup-shaped nest of soft grasses is constructed in a hollow scratched out under the shelter of a small bush, tuft of grass or stone. The two, three or four eggs are pale grey marked with spots of dull brown and grey.

PLOVERS AND DOTTERELS

FAMILY *CHARADRIIDAE*

Although part of the large 'wader' grouping of birds, the plovers do not wade, but live on open grasslands well away from the sea, as does one of the dotterels, the Australian or Inland Dotterel, which inhabits the dry interior. The plovers and dotterels are rather plump birds with short necks but medium to long legs which are strongly developed by their terrestrial way of life. Bills are rather short and straight, and have a characteristic thickening near the tip. Plumage patterns tend to be bold, with contrasting areas of white, black, greys and browns. The plovers of the 'lapwing' group (genus *Vanellus*) commonly have colourful wattles on the face and spurs at the angles of the wing. Included in this group are the Banded, Masked and Spur-winged Plovers. These tend to be noisy, and often aggressive if their nest is approached. Most members of the family readily employ a distraction display, the 'broken wing act', if any intruder comes near their nest or young. This worldwide family contains about fifty-six species, of which seventeen occur in Australia, and six are found only within this region.

Banded Plover   *Vanellus tricolor*

This plover, which is most often seen on the ground in open country in almost any part of the southern half of Australia, has an attractive plumage of black, white and brown. The crown and the side of the head are black, continuing down the neck on either side and meeting as a broad band across the breast. The throat and abdomen are pure white, while the back and wings are brown.

The Banded Plover has become a common bird of pastures and paddocks as land clearing has created greater areas of suitable habitat. Its nest is a slight depression in the ground in which its three or four eggs are laid. These vary in colour from green to brown according to the colour of the soil and vegetation of the paddock.

Spurwing Plover   *Vanellus novaehollandiae*

This inhabitant of the damp grasslands and open country bordering estuaries and swamps is essentially terrestrial in habits, spending most of its time on the ground where it can, like other plovers, run rapidly on long legs. This has not resulted in any loss of powers of flight, the bird taking to the air to avoid danger, and aggressively 'dive-bombing' any intruder near the nest site. After the breeding season flocks form, sometimes very large and seasonal nomadic travels may be undertaken, sometimes over great distances even as far abroad as New Zealand.

The Spurwing Plover derives its name from the sharp spur at the bend of the wing. It is light grey-brown on the upper surfaces, black on the crown of the head extending down the back and sides of the neck to form an incomplete collar, with a black bar across the tail. The entire underparts are white, and there are conspicuous yellow facial wattles. The four grey-green, brown and reddish-spotted eggs are laid in a slight hollow in the ground.

Hooded Dotterel   *Charadrius cucullatus*

This small (175 millimetre length) bird shows strange variations in choice of habitat. In south-eastern Australia the species is restricted to marine beaches, and generally does not venture inland, but Hooded Dotterels in south-western Australia are abundant on inland salt lakes as well as sea coasts and estuaries. The western population is isolated from the east by the desert-like Nullarbor Plains, and the Great Australian Bight, where cliffs plunge to the sea without any beaches.

The Hooded Dotterel is brown on its upper parts, very dark on the head, throat and upper back. The under parts except for the throat are white. A touch of bright colour is provided by the pink legs, the black-tipped orange bill, and scarlet eye-ring.

The nest is merely a slight hollow, but the eggs, of a sandy colour spotted and blotched with blackish and brownish markings, are very well camouflaged.

PITTAS

FAMILY   *PITTIDAE*

The pittas are small birds of almost jewel-like beauty which inhabit the rainforests, jungles, tropical bamboo forests and high altitude wet moss forests of Africa, India, south-east Asia, New Guinea and Australia. Pittas are plump birds with large heads, and tails so short that they often appear to be missing altogether; the legs are long and thin, the feet large. They usually keep to the floor of the rainforests, hopping rapidly rather than flying. The wings are rather short, and rounded, and flights are generally brief except on migrations. These birds feed among the forest debris, turning leaves for insects, worms and snails. Although the twenty-three species have such a general similarity that they are all grouped within a single genera, there is a great diversity of plumage patterns and colours, which are remarkably bright, including strong hues of red, blue and green set amid rich brown and patches of white and black.

Buff-breasted Pitta   *Pitta versicolor*

An inhabitant of the rainforests of coastal north-eastern Australia (as far southwards as the north-eastern corner of New South Wales) the Buff-breasted Pitta is a shy and elusive bird, especially during the breeding season.

The pitta is a terrestrial bird, which searches the leaf-litter of the rainforest floor for insects, and snails which it smashes against rocks.

All of Australia's four species of pittas are colourful birds. The Buff-breasted Pitta is brown on the forehead and crown, and has a broad black band from around the bill and under the chin extending back across the sides of the face (above and below the eye) and around the back of the neck. The breast and flanks are buff, with black down the centre of the belly, and the under-tail coverts are red. The back and wings are bright green, the shoulders and rump are shiny pale blue, and the tail is black.

The Buff-breasted Pitta's nest is a large domed structure of twigs and moss, lined with leaves placed on or very near the ground, usually between the buttress roots of rainforest trees, between fallen logs, among rocks or in a large low fork of a tree. The four eggs are creamy-white, spotted with brown and grey.

PRATINCOLES AND COURSERS

FAMILY   *GLAREOLIDAE*

This family of about 15 species is divided into two main groups, the coursers and the pratincoles. The former are regarded as being the more primitive, feeding on the ground and having only three toes, longer legs, and square-cut tails while the pratincoles have long pointed wings and, usually, deeply forked tails giving them a swallowlike appearance. However the single endemic Australian species is unique in having features of both coursers and pratincoles but is distinctly different from both those groups; it is sometimes known as the Australian Courser. Two members of the family occur in Australia, the other being the Oriental Pratincole, a non-breeding visitor from northern Asia.

Australian Pratincole   *Stiltia isabella*

This small (twenty centimetres) bird of the open plains is rufous-brown on the upper parts, with flight quills and tail black. The throat is greyish white and the upper tail-coverts white (conspicuous in flight). The breast plumage is brown, the abdomen dark brown, and the lower abdomen and under tail-coverts white.

An inhabitant of the dry inland plains, this pratincole is by its colours well camouflaged, so that it is usually only when it runs that it is noticed. It feeds on insects, taken both on the ground and in flight. The two eggs, of a salmon colour spotted with brown and grey, are laid on the ground, often surrounded by a ring of sticks or small stones. This species inhabits most of the Australian continent except for south-east and south-west.

RAILS AND CRAKES

FAMILY   *RALLIDAE*

The rails and crakes are secretive inhabitants of tall grasses and reedbeds in marshy places. They are small to medium sized birds, rather hen-like in general shape. Their bodies tend to be narrow, which enables them to move more easily through dense clumps of reeds and matted grasses. The tail is usually short, and many species have a habit of flicking the tail up and down. The legs are long, and these birds run fast on land (the rails in particular); some are strong swimmers and, like the coots, obtain much of their food by diving. During the daylight hours it is difficult to flush these birds from their reedy haunts, and when forced to fly, it is usually only for a short distance, their flight looking awkward, with long legs dangling down. Some species, however, make long migrations to new areas of suitable habitat, and these flights are usually at night. In most cases the plumage is in sombre colours or streaked and barred to blend with the swampland background, but some have patches of colour, like the red bill and purplish breast of the Swamp hen. Sixteen occur in Australia out of a world total of 132.

Banded Rail   *Rallus phillippensis*

A colourful bird of the swamps and reedbeds, rather hen-like in general shape, with cocked-up tail and strong legs; length thirty centimetres. The plumage is very attractively patterned, the most distinctive feature being bold black bands across the white of the neck, breast, belly and under-tail coverts. There is a patch of bright cinnamon on the upper breast, and through the eye extending to the back of the neck; above the eye is a white 'eyebrow' line. The crown, back, wings and tail are dark olive-brown marked with darker brown, the back and wings speckled with white.

The Banded Rail inhabits dense vegetation bordering swamps and lakes, where it is difficult to see or flush out, being able to run very fast through the reeds or grass and skilfully conceal itself from view. The nest is a shallow scrape on the ground, usually screened by overhanging vegetation. The five or six dull white eggs are spotted with grey or brown.

Black-tailed Native-hen   *Gallinula ventralis*

Rather hen-like in shape, upper parts olive-brown, darkest on the head and black on the tail; throat and breast blue-grey, shading to black on the belly, legs and under-tail. The legs are red, the eyes yellow, the bill green above and on the tip, and reddish below.

The Black-tailed Native-hen has an extremely wide distribution, being absent only from the northern extremities of the Northern Territory and Cape York; in Tasmania it is replaced by the Tasmanian Native-hen. The nomadic native-hen wanders wherever heavy rains have created a favourable habitat, and there are records of mass migrations of huge numbers of these birds. The preferred habitat is the marshy, reedy margins of swamps, lakes and rivers; from the dense low vegetation the native-hens venture out to feed on nearby open damp grasslands.

Marsh Crake   *Porzana pusilla*

This small (fifteen centimetres) inhabitant of the thick vegetation around swamps is very secretive, always moving quickly from one tussock to another, keeping a screen of foliage between itself and any observer; the tail is constantly flicked up and down.

The face and the entire undersurfaces of throat and breast, from bill to legs, are pale grey; from the legs back onto the under-tail is white with transverse black bars. The crown, nape of neck, back, wings and upper tail surfaces are brown streaked with black, spotted with white

on the upper back. The eyes are red and the long legs and long-toed feet are greenish-yellow.

The nest is domed over with interwoven grass stems, and hidden among grass, or in tussocks of vegetation in shallow water or on dry land.

Swamp hen   *Gallinula porphyrio*

An inhabitant of the swamps, where it hides in the dense vegetation of reeds and tall grasses, the Swamp hen is black on the upper parts, and blue-violet from the lower face, neck and breast, and edges of the shoulders, to belly. A large 'shield' extending up onto the forehead from the bill is, together with the bill, bright red. The eyes also are red, while the legs may be red, brownish or grey-green.

The nest is built by bending over and trampling down the reeds or other vegetation to form a platform, where up to five buffy, brown-spotted eggs are laid.

ROLLERS

FAMILY   *CORACIIDAE*

These birds are named for their rolling, diving, twisting display flights, which are accompanied by loud harsh calls. They are solidly built birds with thick hook-tipped bills and, like kingfishers, relatively large heads. They hunt flying insects, dashing out from a perch at the top of a tall forest tree or other high vantage point. Their tails are long, frequently forked; the legs are short and weak. The headquarters of these birds appears to be Africa where many species occur. Of the seventeen species, only one reaches Australia.

Dollar Bird   *Eurystomus orientalis*

This colourful bird attracts attention by its acrobatic flights, when it displays by rolling and tumbling in the air, while uttering loud harsh rapidly repeated 'kuk-kak' calls.

The name is derived from the pale blue coin-like spot on the undersurface of the outstretched wing (the species occurs also in China, India and the islands to the north of Australia, so presumably the name originated where dollar coins rather than notes were in use). This bird, when not catching insects in the air, is usually seen perched on a high bare branch, when it appears to be a greenish bird with a red bill.

Male and female are alike in plumage colours. The head, neck and upper breast are brown shading to turquoise on the back and wings, and to pale olive green. The throat is deep violet streaked with pale blue. The flight feathers are black edged with purple, and the tail deep blue. The feet, bill and ring around the eyes are bright red.

The Dollar Bird is a migratory visitor, arriving from New Guinea and other northern islands in September, breeding in Australia, and departing about April. The nest is an unlined hollow in a branch or treetrunk, almost invariably very high and in a dead tree; the four eggs are white.

SANDPIPERS, SNIPE, CURLEWS, etc.

FAMILY   *SCOLOPACIDAE*

The world's seventy or more species within this family are mostly waders which range from small to medium in size, more slender than the plovers and dotterels, with a smaller head and longer bill. The snipe is an exception to the wader way of life, preferring dense vegetation. All thirty-one members of this family occurring in Australia are migratory summer visitors which breed in the northern hemisphere. Identification of many of these is difficult as the plumage tends to be in greyish tones (brownish colours of the breeding plumage are seen only in the northern hemisphere) and some species have very few obviously characteristic features. There is also a bewildering variety of names — within this family are not only the sandpipers, snipe and curlews, but also various whimbrels, knots, tattlers, stints, and godwits, as well as the Dunlin, Greenshank, Sanderling and Ruff. Bills sometimes aid in identification, for while most are unexceptionally straight mud-probing bills, some are distinctly downcurved (the whimbrels and curlews) or gently upcurved (Terek Sandpiper, Greenshank and Godwits); in other cases behavior such as head-bobbing gives a clue to identification.

Eastern Curlew
Numenius madagascariensis

The largest of the summer migrants from the northern hemisphere, the Eastern Curlew is fifty to sixty centimetres in length, of which the long downcurved bill makes up about sixteen to eighteen centimetres. The plumage is rather indistinctive, being principally brownish with lighter and darker streaks. The face, front of neck and remainder of underparts are buffy-white streaked with grey-brown. The flight feathers are almost black and the tail is barred with very dark brown.

The Eastern Curlew favours muddy beaches where it probes with its very long bill for small marine creatures.

Japanese Snipe   *Gallinago hardwickii*

A small bird of the swamps and marshy grasslands, this snipe is one of two species visiting Australia from the northern hemisphere (the other being the Pintailed Snipe). It has a very long, perfectly straight bill, the crown of its head is black with conspicuous buffy streaks, and the remainder of the upper parts is buffy-brown patterned with black. From the face down onto the breast is pale cinnamon brown streaked with darker brown, and the remainder of the underparts are white. The tail is a bright reddish-brown.

The Japanese Snipe frequents the margins of lakes, swamps and places where there is damp grassland or other moist vegetation. Its flight when it is disturbed is swift and erratic, and after landing it immediately runs very fast for a short distance.

Sharp-tailed Sandpiper   *Calidris acuminata*

An inhabitant of the coastal and inland mudflats and sandbars around most parts of the Australian coast, the Sharp-tailed Sandpiper is one of the many migratory waders which breed in north-eastern Siberia and spend part of the year in Australia.

These small (twenty centimetre length) sandpipers are streaked grey, brown and black on the upper surfaces, and have the rump and tail black in the centre and white on the sides. There are rufous and black streaks on the crown, while the breast is white with buff and brown markings. The bill is black and legs yellowish to olive-green. The green legs and the pattern of the rump distinguish this common bird from the rather similar Curlew Sandpiper, with which it frequently associates on the beaches.

Sharp-tailed Sandpipers feed on the usual wader diet of insects and other small aquatic creatures.

Turnstone   *Arenaria interpres*

This small bird of the coasts is a visitor from the northern hemisphere. It is seen here only in its non-breeding plumage, when the upperparts are dark chocolate brown marked with lighter brown and black. The throat is buffy-white, the breast brownish-black, belly white and legs orange. In flight a broad white bar is visible on the outstretched wings.

The Turnstone inhabits rocky shores and exposed reefs, running swiftly and stopping every now and then to turn over a stone, shell or seaweed debris to reveal small crustacea and insects.

Whimbrel   *Numenius phaeopus*

One of the most distinctive features of the Whimbrel is its extremely long, slender down-curved bill, which accounts for more than seven centimetres of the bird's total length of forty to forty-five centimetres. Often features which help identify this wader are the grey and black streaks on the crown, the white streak above each eye, the dark grey-brown and white mottled back, the grey-barred, whitish rump and upper tail, white chin, white breast and belly, and tail that is barred with dark grey-brown.

This summer visitor from the northern hemisphere does not breed within Australia. It frequents the mudflats, wet grasslands, coastal marshes and estuaries, and often feeds on areas exposed by low tide. The Whimbrel may be seen in any Australian coastal locality between July, and April.

SCRUB-BIRDS

FAMILY   *ATRICHORNITHIDAE*

This family is represented by only two species, one of which is uncommon and of localized distribution; the other is extremely rare and confined to one very restricted locality. This family is unique to Australia, and is particularly interesting because in anatomy, song, ventriloquial powers and nesting habits the scrub-birds appear most closely related to the far bigger lyrebirds. They may represent a primitive avarian stock which has survived only in these widely separated pockets of suitable habitat. The Noisy Scrub-bird in particular seems to be on the verge of extinction, and recent studies suggest that it is most exacting in its environmental requirements. Scrub-birds have short rounded wings and, although not flightless, prefer to keep to the ground under cover of very dense vegetation. Plumages are rather dull browns relieved by some lighter patches.

Noisy Scrub-birds   *Atrichornis clamosus*

One of Australia's rarest birds, with a very small population confined to an extremely restricted area of coastal hills and scrubby heathlands in south-western Australia. For many years the Noisy Scrub-bird was thought to be extinct; it was accidently rediscovered only in 1961, having been 'lost' since 1889.

This bird and the closely related Rufous Scrub-bird (which inhabits mountain top rainforests of Queensland and New South Wales) are probably among the most primitive of songbirds; anatomically they are considered to have a close affinity to those other famous songsters, the lyrebirds. The scrub-bird family, of which these are the only species, is unique to Australia.

The Noisy Scrub-bird is dark brown with white streaks on each side of the chin. The tail is long while the wings are short and rounded and the legs and feet large and strong. The bird spends most of its time on the ground under dense low scrub. The call is a loud penetrating 'chip-chip' and it often imitates the calls of other birds.

SILVEREYES

FAMILY   *ZOSTEROPIDAE*

The silvereyes, white-eyes or greenies are small (12-15 cm), mainly green-plumaged birds which inhabit forests from Africa through India, China, Indonesia and New Guinea to Australia and New Zealand. Their distinctive feature is a prominent ring of white or silvery feathers around each eye. (A few species, not Australian, have the eye ring of some other colour, or missing). Silvereyes, rather honeyeater-like in habits, feeding on nectar, fruits and insects, are of uncertain origins, also resembling in various respects the sunbirds and the flowerpeckers. Their bills are rather short and almost straight, the tongues brush-tipped, and they feed upon the ground as well as in trees and shrubs. They peck holes in the corollas of some deeply tubular flowers where their short bills would not otherwise reach the nectar. Nests in this family are small baskets woven of grass, bark fibres and spiders' webs, hanging by their rims from horizontal forks in the outer foliage of trees or shrubs. Four of the total of eighty species occur here and two are found only in Australia.

Western Silvereye   *Zosterops gouldii*

This Silvereye, which has its entire upper parts dark grey-green instead of the grey back, green crown and rump of the Eastern Silvereye, is confined to the south-western corner of the continent. In that region it is probably the commonest of all small birds, large flocks forming in autumn and winter. Silvereyes forage in suburban gardens around Perth, where they take the place of the Sparrow of the eastern cities, that unwanted introduced species never having become established in Western Australia.

Although the Silvereyes through their abundance, cause some damage to grapes and other fruits, they also consume enormous quantities of insects. The nest, a suspended cup-shaped structure of grasses, wool, and cobwebs, is suspended among foliage of a bush or tree, often in garden shrubs.

THE SITTELLAS

FAMILY   *NEOSITTIDAE*

Five of the total seven species of sittellas occur in Australia. They are small birds, less than about twelve centimetres in length, with short tails, large powerful feet and sturdy legs, for they move in any direction, up, down, sideways, upside down, as they hop across the bark of trees pecking at insects in the crevices of the bark. Their bills are straight but quite long and robust, and appear to be set at a slightly upwards angle from the head, no doubt in some way best suited to their bark-probing habits. In flight sittellas are easily recognizable by the conspicuous orange or white wing patches. The sittellas are widely distributed throughout eucalypt forests and woodlands, but although there are five species, there is little overlap of distribution; they are probably geographical races of the one species which are almost sufficiently distinct from one another to be distinct, and therefore are generally regarded as full separate species.

Black-capped Sittella   *Neositta pileata*

Sittellas forage on the bark of trees, spiralling around and down each limb and trunk until near the ground, then flying to the top of the next tree to repeat the process. They move in quick jerky bursts, clinging as easily to the undersides of the branches as to the upper surfaces. Often they are in small family parties of around half a dozen birds, even during the breeding season, when all in the party join in feeding the young.

The nest is a masterpiece of disguise. It is blended into a vertical fork, often of a dead limb, and usually quite high. It is a beautifully constructed nest, deeply cup-shaped, using spider's webs and cocoons to bind the materials to a tough flexible felt-like consistency. The base is blended onto the limb or sides of the fork so that it looks like a broken end of a branch, an effect heightened by the use of small flakes of bark from the same tree bound on as the outer layer. The cunning behavior of the birds makes the nest much harder to locate. When approaching the nest with food in their bills they begin at the top of the tree and spiral down the bark exactly as if searching for insects, pop the food into the beak of the young, and continue down the branch before flying off. This sittella occurs over most of Australia except the north and east where it is replaced by similar species.

SKUAS

FAMILY *STERCORARIIDAE*

The skuas are gull-sized marine birds of prey which inhabit the seas and coasts around the world, taking on the same role over that habitat as have the hawks, falcons and eagles over the land. However they are not related, being aberrant gulls which have become adapted for the predatory role, and acquiring swift flight, extreme aggressiveness, long pointed wings, and talons strong and curved like those of an eagle. In flight they are hawk-like; the head also is like the land birds of prey having a fleshy cere at the base of the bill, which is sharply hooked. These characteristics have been developed through the process of convergent evolution, both the land birds of prey and the skuas having so similar a way of life. Most powerful is the Great Skua, which is the terror of the colonies of penguins and other birds in the Antarctic region. These birds are said to be more pugnacious than any falcon, more courageous than any other bird in defence of their nests. There are four species of skua, all frequenting Australian coasts.

Great Skua   *Stercorarius skua*

This visitor to Australia's southern coasts (as far north as Queensland's Fraser Island and the Abrolhos Islands in Western Australia) breeds in the Antarctic, and at other seasons wanders the southern oceans. In common with the three other species of skuas visiting these shores, the Great Skua is a 'pirate of the seas'. Very large (wingspan 2.5 metres) and armed with a massive hooked bill, the skua harrasses other sea birds which have been fishing, forcing them to drop or disgorge their load in an attempt to escape; the skua then swoops on the fish. Skuas also prey upon unguarded young in seabird colonies, and fulfil a scavenging role on coasts and oceans.

The Great Skua is dark grey-brown, paler beneath. There is a large white patch on the underside of each wing, visible from below on the flying bird.

STARLINGS

FAMILY *STURNIDAE*

This large and almost world-wide family is made up of birds which typically are highly gregarious; they are very adaptable (and hence extremely successful colonists of new regions). Starlings are usually wholly or largely black plumaged, the black having a glossy iridescent sheen. Most starlings nest in cavities of trees, walls or cliffs, but others, like the Shining Starling, build a massive communal nest containing many chambers. In this family parental care is rather primitive, both sexes sharing equally in all duties of nest-building, brooding and feeding of the young. Some ornithologists consider the Shining Starlings and others of their genus to be the most primitive of the family. Of the one hundred and ten species, Australia has only one indigenous and two introduced starlings.

Shining Starlings   *Aplonis metallica*

Found only in the rainforests of tropical north-eastern Queensland (outside Australia it occurs also in New Guinea) the Shining Starling is our only native species of starling. Unlike the two introduced members of this family, it is a most attractive bird. The general colour of the plumage is black, but highly glossy, reflecting the light with purple and green highlights. The length averages twenty-five centimetres of which about eleven centimetres is taken up by the sharply tapering tail. The eyes are brilliant red and set in black plumage showing up like glowing coals in the dark.

Shining Starlings are arboreal in habits, feeding mostly in the treetops but occasionally coming to the ground to find insects and fruits. These birds keep together in flocks, and are colonial nesters. A great many bulky nests are suspended from the branches of a single tree. They are domed or spherical structures roughly woven using vines and tendrils, and lined with softer materials. This species is a migratory visitor to Australia from New Guinea, coming here to breed between August and May.

STONE CURLEWS

FAMILY *BURHINIDAE*

The nine species (two within Australia) of this family resemble small bustards, and, like them, some have left the shores and inhabit stony semi-desert country far inland. The stone curlews have discovered a profitable niche that has been left unoccupied by other birds in most of the warmer parts of the world — they feed at night, finding insects, worms, small lizards, rodents and molluscs. During the day they sit with their legs folded under them, and necks stretched out on the ground. With their cryptic plumage colours

and patterns they are then extremely difficult t detect, and have such great confidence in thei camouflage that they can be very closel approached and even touched without moving These birds have bulging knees and sometime are known as 'thick-knees' as well as 'ston plovers' and 'stone curlews'. One of the Aus tralian species is endemic.

Bush Curlew   *Burhinus magnirostris*

Occurring in open woodlands and plains countr throughout Australia, the Bush Curlew is a noc turnal bird whose loud mournful 'ker-loo' make a haunting sound at night in the bush. During th day the bird is seldom seen, hiding in a patch o vegetation and 'sitting tight' relying upon camou flage even when closely approached. It does no readily fly, at least during the day, but runs fas with head down. This species appears to hav become less common following the introductio of cats and foxes to Australia.

The Bush Curlew's eyes, yellow and ver large, indicate the nocturnal habits of this bir which is about fifty-five centimetres in length and has grey-brown plumage marked with black on the back; there is a white shoulder patch, an the breast is buffy with darker streaks.

The Curlew's nest is no more than a sligh depression in the ground, usually partly con cealed by a bush or other vegetation. Th colours of the eggs vary to match the ground, bu are usually greyish or brownish with darke markings.

STORKS

FAMILY *CICONIIDAE*

Massive bills, long legs and necks, a slow, statel walk, and powerful flight with neck and leg extended, are features characteristic of storks The only other birds with which they might be confused are the cranes (in Australia the Brolg and Sarus Crane) which however are more slender, lighter and graceful, with fine bills and grey in plumage. Storks can cover great dis tances on their migratory travels; their long and broad wings giving them soaring and gliding capabilities. Of the world total of seventeen species only one species reaches Australia.

Jabiru   *Xenorhynchus asiaticus*

A very large bird (standing 120 centimetres tall) with a very upright posture; the bill is very long (thirty centimetres) and heavily built. This is Australia's only stork, and its general appearance is unmistakeably stork-like. The Jabiru is an impressive bird not only in its size but also for its plumage. Its head and neck are black with an iridescent dark green sheen, becoming glossy purple and bronze on the nape of the neck. The wings have a broad black band on the shoulders, the tail and bill are black, while the rest of the body is pure white. The very long legs are deep pink.

The Jabiru is usually seen in or near water, in coastal or inland swamps, river pools, estuaries and lagoons. Its distribution covers northern tropical Australia and eastern Australia excluding the arid interior. The nest is a rough platform of sticks placed high up in a tree.

STORM-PETRELS

FAMILY *HYDROBATIDAE*

The small swallow-like storm-petrels (sometimes known as Mother Carey's Chickens) differ anatomically from the generally larger petrels, shearwaters and prions in that the nasal tube is undivided. Their legs are long (for sea birds) and slender, for they often flutter just above the surface with feet pattering over the surface, snatching up surface plankton. Only one species, the White-faced Storm-petrel, breeds on Australian shores.

White-faced Storm-petrel
Pelagodroma marina

Of the seven storm-petrels visiting Australian seas the White-faced is the only species to breed here; it is also the only member of the family which is entirely white on the underparts. Also white are the forehead and face (where there is a contrasting black patch around the eye) and the rump. The upper parts are dark grey-brown, darker on the crown, tail and wings.

The White-faced Storm-petrel wanders the open oceans of tropical and sub-antarctic regions. It keeps low over the water, fluttering above the wave-tops with legs dangling down, dropping down to feed on surface plankton.

Nesting occurs on many islands offshore from the coastlines of New South Wales, Victoria (where a colony occupies some 8,000 burrows on Mud Island in Port Phillip Bay), Tasmania, South Australia and Western Australia.

SUNBIRDS

FAMILY *NECTARINIIDAE*

The sunbirds of the Old World, from Africa to India, south-east Asia and Australia, are the ecological counterparts of the hummingbirds of the Americas. But although similar in appearance and habits, they belong to two separate orders of birds, sunbirds being small perching birds and the hummingbirds being related to the swifts. Like hummingbirds, sunbirds often hover in front of a flower with wings beating so fast that they are but a blur, probing for nectar with long down-curved bill and slender tubular tongue. More often they perch to feed at flowers, or to search for small insects and spiders which may be found by hovering beneath foliage while scanning the undersurfaces of the leaves. Large pendant nests with side entrances are the general pattern of construction in this family. Sunbirds are renowned for their brilliant plumages, which often include iridescent colours.

Yellow-breasted Sunbird
Nectarinia jugularis

A honeyeater-like bird of the rainforest margins, this sunbird (the only species occurring in Australia) is usually seen darting among the foliage, sometimes catching insects in flight, hovering briefly in front of a flower to drink its nectar, or searching among the leaves. Although resembling other Australian honeyeaters, the sunbird is a member of quite a different family. Likenesses, such as the very long, slender, downcurved bill result from its similar nectar-seeking way of life.

The male Yellow-breasted Sunbird is 12 centimetres in length, with the slender bill making up about 2 centimetres of this. His upper plumage is olive, dark brown on the wings, and on the tail black, tipped with white on the outer feathers. His throat and upper breast are deep iridescent purplish-black, changing abruptly to bright yellow for the remainder of the undersurfaces.

The female is similar but lacks the iridescent black throat patch, being entirely yellow beneath.

The sunbird's nest is suspended by its roof (being a domed structure) from a twig of a bush or low tree, sometimes from a convenient part of a building. The lower part of the nest terminates in a very long hanging 'tail', while the entrance is a small hole in one side near the top. It is constructed of various soft plant fibres bound together with cobwebs; the two or three eggs are greenish with markings of dark brown.

SWANS, GEESE AND DUCKS

FAMILY *ANATIDAE*

The major groups within this family — swans, geese, and ducks — although superficially distinct, are in many ways alike in anatomy and behaviour. All waterfowl have very short legs, and a quite long straight bill with laminations along the edges. Most food is taken underwater, the water and mud in the bill being sieved out through the ridges when the head is raised. Some geese have the laminations modified for shearing of the grass upon which they graze. Waterfowl have ten primary feathers in the wings, but a variable number of tail feathers. During the

moulting of the primaries, usually after the breeding season, many waterfowl become flightless for a period of weeks. All members of the family Anatidae have very dense waterproof plumage with a heavy underlayer of down, which they pluck out for the lining of their nests. Large clutches of eggs are normal, usually more than ten, and up to twenty; the downy young are able to swim as soon as hatched. Waterfowl are gregarious, and usually feed and travel in flocks, sometimes of great size. The family is divided into three main sections. One consisting of a single species, the Australian Pied Goose which is a unique primitive form of goose; the second containing other geese, swans, and the whistle-ducks; the third group or sub-family consists of the great variety of ducks. Altogether there are 147 species in the family, twenty-three occurring in Australia, of which ten are found nowhere else.

Black Duck   *Anas superciliosa*

The Black Duck is a member of a group of ducks known as 'dabbling', that is, surface-feeding, ducks. The ducks of this world-wide genus of about thirty-five species feed in shallow water by tipping tail up, beak down, rather than submerge fully and dive to the bottom of deeper water, as do the various species of diving ducks.

The Black Duck occurs throughout Australia and Tasmania wherever there is suitable habitat of fresh, brackish and sometimes salt water wetlands, lakes or rivers; however it visits the more arid parts of the interior after heavy rains.

Distinctive features of this duck's plumage are the dark crown, white and black facial streaks, buffy-white edging to the dark brown body feathers, and dark green iridescent, black-bordered wing speculum.

Like many other Australian birds, the Black Duck varies its breeding season according to rainfall. The nest may be a hollow in a tree or a slight depression in the ground among dense vegetation. The eight to ten eggs are almost hidden beneath the soft down which is used by the duck to line her nest.

Black Swan   *Cygnus atratus*

Entirely black except for the white flight quills, red beak and red iris to the eye, this swan is one of the most famed of Australian birds. It may be seen in any part of Australia, but much more commonly in the south-east and south-west,

where it occurs on any water habitat particularly fresh or brackish swamps and lakes.

The Black Swan builds a bulky nest of sticks and rushes which may be on a small island, built into a bush growing in water, or floating among rushes. The nest itself is almost a miniature island, especially when of the floating type, being a metre or more in diameter. The four to eight eggs measure about 100 by 67 millimetres, have an average weight of 247 grams, dull-surfaced and very pale green.

Cape Barren Goose
Cereopsis novaehollandiae

This unique Australian bird appears to have no close relations among present-day birds; it probably belongs to the duck sub-family and may be a primitive link between this and the goose sub-family. It is confined to the islands of the southern coasts of Australia from Cape Leeuwin in Western Australia to the New South Wales — Victoria border in the east; in parts of its range it visits mainland Australia.

The Cape Barren Goose is very large and solidly built. It is ashy-grey, with scattered black spots over the shoulders. The tail is black above, the tips of the flight feathers are black, and the crown of the head is white.

This goose is very rarely seen on water. It feeds on grasslands, eating grass and other vegetation, and may be seen in small or large flocks. The nest, on the ground or in dense thickets, is made of grass or other vegetation, and lined with down. The four or five eggs are white.

Grass Whistle-duck   *Dendrocygna eytoni*

A duck of distinctive appearance due to its upright, goose-like carriage, and the fact that the flank plumes are held upright outside the folded wings. The upper parts are brownish-black, the rump and tail dark brown, wings brown, abdomen buff, breast chestnut barred with black, and the long lanceolate plumes on the flanks are edged with black. This species is also known as the Plumed Tree-duck.

In flight these ducks whistle loudly and incessantly. They inhabit tropical grasslands usually near water, but feed on the grassy

surrounds rather than in the water. Grass Whistle-ducks, are distributed across northern Australia from the Kimberleys to Queensland and in eastern Australia extend southwards as far as the vicinity of the Murray River.

Grey Teal Duck   *Anas gibberifrons*

This duck, which may be found in any part of Australia or Tasmania, shows remarkable behavioural adaptations for the semi-desert conditions which prevail over much of the continent.

The Grey Teal will be absent from arid regions at times of drought, when all surface water has dried out. But where rain does occur, filling the thousands of shallow salt lakes, claypans and waterholes, this is one of the first birds to fly in (from the more permanent coastal wetlands) to make use of new feeding grounds. It is one of the most successful opportunistic breeders among Australian waterfowl — nesting will occur at any time of the year whenever rainfall and rising water levels make conditions suitable.

Grey Teals wander far in their search for optimum environmental conditions; no other Australian duck travels so extensively. There appears to be no regular pattern of migration, but rather a general scattering outwards from unfavourable areas, and concentration at areas which have experienced heavy rains or floods.

The Grey Teal is principally grey-brown with white throat and light markings; there is an iridescent wing speculum of green and white.

Water Whistle-duck   *Dendrocygna arcuata*

Resembles the Grass Whistle-duck, but generally darker in plumage, with the entire upper surface including the top of the head blackish-brown. The primary wing feathers and outer wing-coverts are black, and the innerwing coverts deep chestnut. The breast is rufous, the abdomen chestnut and the long flank plumes white with broad chestnut edgings. At a distance it can be noticed that this species stands with a more horizontal body posture than the erect-standing Grass Whistle-duck; it often feeds in water by diving, which the Grass Whistle-duck rarely if ever does.

The Water Whistle-duck is restricted to a narrow belt of suitable wetland country, of river floodplains, lagoons and swamps, along Australia's northern and north-eastern coasts; it also occurs in New Guinea, Indonesia and the Philippines.

SWALLOWS AND MARTINS

FAMILY   *HIRUNDINIDAE*

Swallows and martins are graceful small birds which obtain all their insect food on the wing, diving, soaring and twisting effortlessly to the accompaniment of high-pitched twittering sounds, usually in small colonies or flocks. The superb powers of flight of the swallows are facilitated by the exceptionally long wing primary feathers. The feet are so small that they can only shuffle awkwardly across the ground, but serve adequately for perching on thin twigs. Swallows and martins (these names are more or less synonymous) are small, the Australian species all being less than 18 cm including the long forked tails. The plumages of these birds are usually iridescent black or dark above, and white with reddish areas below, and sometimes white patches on back or rump. Their short bills have an extremely wide gape, and are used for scooping up swarms of tiny flying insects. Although some species are sedentary, most are nomadic, and some undertake regular migrations. At least one species, the White-backed Swallow, is known to reduce its metabolic rate and become torpid, when insects are scarce. Although these birds resemble swifts, they are less perfectly adapted to the aerial life, wings are not as long and curved, and they are not as streamlined.

Swallows and swifts are not closely related; the similarity of appearance results from the similar way of life. Three species of this family are unique to Australia, and five of the total seventy-five species occur here.

Fairy Martin   *Petrochelidon ariel*

Black on its upper surfaces except for the head, which is dark rusty red, and the rump, which is dull white. The wings and the shallow-forked tail are dull black. Underparts are mainly white, with a touch of grey on the throat, and the tinge of brown under the wings and onto the flanks.

The Fairy Martin has a preference for open country, usually not far from water such as river pools, where it will often be seen swooping low over the surface gathering up tiny insects which are so abundant just above still water. At other times it may chase high-flying insects.

Fairy Martins nest in small colonies, the nests, clustered together, are built of thousands of small pellets of mud stuck to the roof of a cave, overhanging river bank, bridge undersurface or other similar location not far from the source of mud. Each is bottle or retort-shaped structure consisting of a bowl about 15 cm. in diameter (forming the base of the nest, the cave ceiling being the 'roof') and a long spout-like entrance tunnel. The bottom of the bowl is warmly lined with grass and feathers.

This species occurs almost throughout Australia except in heavily forested and mountainous areas.

White-backed Swallow
Cheramoeca leucosternum

This swallow has a deeply forked tail. The tail and wings are black, and afford a strong contrast to the pure white of the back; the head is brown mottled with white. Beneath, the throat and chest are white, the abdomen black.

The White-backed swallow, which like others of its family catches small insects in the air, inhabits almost the entire Australian continent except the tropical far north and north-east, the extreme south-west and south-east corners and Tasmania. This swallow is the sole species of the genus *Cheramoeca*, and is unique in having the ability to become torpid in very cold conditions, thus conserving energy.

The nest is a tunnel about 2.5 centimetres in diameter, drilled into a sandy bank, such as a river bank, road cutting, or similar site. In the nesting chamber at the end of the tunnel is constructed a shallow saucer-shaped nest of grass, fine rootlets and leaves. The four or five eggs are pure white.

SWIFTS

FAMILY *APODIDAE*

This family is poorly represented in Australia, five of the total of seventy-six species being recorded here, but only one of these, the Grey Swiftlet, is resident, the other four being non-breeding visitors from the northern hemisphere or from the tropical islands to the north of Australia.

Swifts are related to nightjars and hummingbirds, not to the swallows which they superficially resemble. Insects caught in flight are the sole source of food. The swifts are believed to have the fastest flight among small birds, yet it is at the same time erratic, with glides, rapid changes in course, and long steep dives. They have no rivals among small birds for sustained flight, and must have devised ways of resting in the air. Swifts are usually dark in color, with some patches of white or other light colour especially on throat, rump or belly. The wings are very long, usually extending far beyond the tail tip when in the folded position, and are made up of ten very elongated strong primary feathers and not more than nine small secondaries. The tails take many forms, deeply forked, or truncate, soft or spine-tipped.

Grey Swiftlet *Collocalia spodiopygia*

Perhaps more than any other birds, swifts and the slightly smaller swiftlets are completely adapted to an aerial life. They are capable of very fast flight, and, living entirely upon insects captured in the air, are unrivalled in the bird world for their sustained flight — they probably spend half their waking life in the air. Australia has five species, the Spine-tailed Swift, the Fork-tailed Swift, the Uniform Swiftlet, the Glossy

Swiftlet and the Grey Swiftlet. This last species is the only member of the family to nest in Australia. It is confined to rocky coastal range country and the offshore hilly islands of north-eastern Queensland. Here the nests are built under the shelter of overhanging walls of gorges hidden in the rainforests, or in crevices and caves between boulders. These nests, unlike any other of Australian birds, are made of *Casuarina* needles cemented with saliva onto the rockface.

THE THRUSHES

FAMILY *TURDIDAE*

Although containing about three hundred species, and of world-wide distribution, this family has only three species native to Australia, these being the Southern Scrub Robin, the Northern Scrub Robin, and the Ground Thrush. The two species of scrub robins are thought by some ornithologists to be primitive members of this family, while others believe them to be members of the babbler family. They may have evolved from some ancestral thrush-like birds, and certainly have been in Australia for a great length of time. Only the Ground Thrush, of Australian birds, closely resembles the true thrushes, and it is probably a comparatively new arrival to this continent. The thrushes are mainly terrestrial birds, feeding among ground litter and nesting near or on the ground.

Ground Thrush *Zoothera dauma*

An inhabitant of the dense forests, particularly rainforests and damp gullies, of eastern and south-eastern Australia and Tasmania. A separate, very slightly different race occurs on the Atherton Tableland. Male and female are alike in plumage, with upper parts light brown, sometimes with white on the inner edge of the outer tail feathers. The face and underparts are basically white, but most feathers have a black margin to their rounded outer end, giving a scalloped or scaly appearance, most evident on the breast. There are black streaks from the base of the bill to below the eye.

The Ground Thrush, as the name implies, is mainly terrestrial. It feeds on the floor of the rainforest, raking aside the layer of fallen twigs and leaves for insects and other small invertebrates. The Thrush's colours blend perfectly with the browns and greys of the forest debris, so that this bird will often be heard scratching about before it is seen. So confident is it of its camouflage that it can often be approached quite closely.

Southern Scrub Robin
Drymodes brunneopygia

Australia has two species of scrub robins. The Southern is unique to Australia, while the Northern Scrub Robin occurs also in New Guinea.

The Southern Scrub Robin is one of the characteristic inhabitants of the inland mallee eucalypt country and drier sandplain scrublands. It seldom flies, but hops quickly across the ground, usually contriving to keep thickets of

vegetation between itself and the observer; it is an unobtrusive bird which can easily be overlooked, and shy enough to be difficult to watch. The food consists of insects and snails obtained under leaf and twig debris.

Usually this scrub robin carries its tail partly cocked up like a wren. Its colours make it inconspicuous. The upper parts are brown, the long tail is reddish-brown tipped with white on the outer feathers. The wings are very dark brown with two white bars on the shoulders and a concealed white bar on the flight feathers. The underparts are buffy-brown. The nest is built on the ground in thick undergrowth. A small hollow is lined with grass bark or fine rootlets. The single egg is greenish-grey with brown spots and blotches.

TROPIC-BIRDS

FAMILY *PHAETHONTIDAE*

Birds of the tropical seas which fish by plunge-diving, these are medium-sized, with the wings and tail smallish for the size of the body — they are not gliders like the albatrosses and petrels, but generally fly straight and high above the water. All three species (of which two visit Australian seas) are mainly white, and have the central tail feathers greatly lengthened, these being white on one species, red on the other. The tail streamers of the White-tailed Tropic-bird are longer than those of the Red-tailed. Their legs and feet are very small, as they do not often come to land except when breeding.

Red-tailed Tropic-bird
Phaethon rubricauda

Of the world's three species of tropic-birds two occur in tropical and subtropical seas off Australian coasts; one, the Red-tailed, breeds in the Australian area. The tropic-bird feeds at sea, diving into the water from considerable heights to take fish and squid. The single egg is laid on the ground under the shelter of overhanging rocks or bushes.

WARBLERS

FAMILY *SYLVIIDAE*

The warblers are principally insect-eaters which capture their food by searching leaves and crevices rather than in the air. This way of life is reflected in the small size of most species (usually less than 15 cm), and the thin pointed bills for pecking at the small insects of the foliage. Plumages are mostly of subdued browns and greys relieved occasionally by patches of brighter colour such as the yellow breast and abdomen of the White-throated Warbler and the yellow rump of the Yellow-tailed Thornbill. Many species have a loud, clear and attractive song, and most build domed nests with side entrance openings. The family is a large one, and in Australia contains the reed warblers (genus *Acrocephalus*), the cisticolas or fantail warblers (*Cisticola*), the grassbirds or marsh warblers (*Megalurus*), the songlarks (*Cinclorhamphus*), the Spinifex Bird (*Eremiornis*), the bristle-birds (*Dasyornis*), the Australian warblers or fairy warblers (*Gerygone*), the Weebill (*Smicornis*), the thornbills (*Acanthiza*), the white-faces (*Aphelocephala*), the scrub-wrens (*Sericornis*), the Fern-wren (*Oreoscopus*), the Red-throat (*Pyrrholaemus*), the Field-wren (*Calamanthus*), the Speckled Warbler (*Chthonicola*), the Rock Warbler (*Origma*), and the Pilot-bird (*Pycnoptilus*). Of the almost world-wide total of about 360 species, 52 occur in Australia, of these, only nine extend beyond Australia.

Rufous Bristlebird   *Dasyornis broadbenti*

The Rufous Bristlebird has become very rare in south-western Australia but remains common in restricted areas of south-eastern Australia. Of its total length of 24 centimetres the broad tail accounts for about half. The head is a bright rufous-brown, the rump chestnut-brown, the wings cinnamon-brown and the remainder of the upper surface grey-brown. The throat and breast have a speckled appearance due to each feather having a black centre and a pale grey edging; the abdomen is grey-brown. The wings are short and rounded.

This bristlebird, one of three species, inhabits dense coastal thickets, where it scratches around among the leaf litter for insects and seeds. It is reluctant to fly, but runs fast, and usually carries the long tail partly cocked upwards and occasionally fanned out.

The nest is a large oval structure of coarse grass, built close to the ground. It has a side entrance, and is neatly lined with finer grasses. The two eggs are dull pinkish-white with numerous spots of brown, grey and amber.

Tawny Grassbird
Megalurus timoriensis

An inhabitant of thick moist grasses, wet heaths, shallow coastal and inland swamps, the Tawny Grassbird is distributed across tropical northern Australia from the Kimberleys in the west through the 'top end' of the Northern Territory to Cape York and down the eastern seaboard into south-eastern Australia. It is on the upper parts cinnamon brown with black streaks, except for the plain rufous forehead, crown and nape. The under parts are almost white.

In the low dense undergrowth which it inhabits the grassbird keeps low and is not often seen except in the breeding season when it often perches on top of a small bush or clump of grass, from which vantage points it makes display song-flights. These birds undertake nomadic travels as seasonal conditions cause drying out and later re-filling of their wet grass and swampy territories. The nest is well hidden in a very dense clump of vegetation, and takes the form of a deep cup-shape structure of grass. The three eggs are pale pink spotted with grey and brown.

FAIRY WRENS, EMU-WRENS, AND GRASS-WRENS

FAMILY *MALURIDAE*

The best-known members of this family are the fairy-wrens (genus *Malurus*), which are small birds with long tails carried cocked upright over their backs. Many species have blue in the plumage, and the males of most are very colourful in breeding plumage. Females, immature birds, young males (which may have begun to breed) and mature males out of the breeding season have nondescript brownish plumages. Wrens of this genus live in small family groups in which various brown-plumaged birds greatly outnumber the males in colourful plumage. Two other types of wrens, similar in size and in their habit of carrying their long tails vertically upright, are included in this family. The emu-wrens (genus *Stipiturus*), among the smallest of Australian birds, have extremely long delicate tails and brownish plumages (the males with bright blue on the throat); they live in pairs rather than the larger social groupings of the blue wrens. The grass-wrens (genus *Amytornis*), resemble the blue wrens in size and shape but are brownish with white or black streakings. They inhabit arid or rough sandstone and canegrass country of the inland and tropics, and like the blue and emu-wrens, spend most of their time on or near the ground. Only one of the total twenty-four species occurs outside Australia, having reached New Guinea.

Banded Wren   *Malurus splendens*

Plumaged entirely in blue, ranging from pale silvery and turquoise blues to deep glossy purplish blue, broken only by bands of black, the male Banded or Splendid Wren, particularly when seen in full sunlight, is one of the most intensely coloured of Australia's small birds. Very similar is the Turquoise Wren, which has black instead of blue-violet on the lower back, and may be a subspecies of the Banded Wren rather than a separate species. The Banded Wren inhabits forests, woodlands and arid scrub country of the southern half of Western Australia, usually where there are areas of dense undergrowth, while the range of the Turquoise Wren continues eastwards from that of the Banded, into the southern parts of the Northern Territory and through central South Australia to Spencer's Gulf.

The Banded Wren, like others of the thirteen Australian species of 'fairy-wrens', form social groups, family parties, so that it is not uncommon to see as many as three males in full blue breeding plumage, together with three or four of the dull grey-and-white females, all combining forces to feed the young at a nest.

Red-winged Wren   *Malurus elegans*

Confined to the heavy forests of Karri and Jarrah eucalypts, and favouring the dense undergrowth of the valleys, swamps and creekbanks, the attractive Red-winged Wren has vanished from areas cleared of natural vegetation, but remains a common species in extensive tracts of forestry,

water catchment and natural park reservations. It occurs only in the extreme south-western corner of the continent.

Although very similar in plumage patterns and colours to the Variegated Wren, Purple-backed Wren and Blue-breasted Wren, the Red-winged Wren can also be separated in its choice of habitat, the other Chestnut-shouldered Wrens being, in Western Australia at least, inhabitants of much drier open forests and arid scrublands.

Red-winged Wrens built a domed nest with a side entrance, using strips of bark, dead leaves and grass. It is extremely well hidden in dense undergrowth, a favourite site being within the dense 'skirt' of dead leaves of a low grass-tree (blackboy) within ten to twenty centimetres of the ground. Extensive spring-time burning-off operations in these forests which are the major habitat of the Red-winged Wrens must therefore cause heavy losses to this species.

Southern Emu-wren   *Stipiturus malachurus*

Among the smallest of Australian birds, the three species of emu-wrens all have extremely long tails made up of six delicate plumes which are of a filamentous structure resembling emu feathers.

The Southern Emu-wren has a slender grey tail which makes up about thirteen centimetres of the bird's total length of about twenty-one centimetres. The upper parts of the body are streaked with grey and black, lighter on the head. The throat, chest, and a line from the bill to just above the eye are pale blue. The ear coverts are streaked grey and white, and the remainder of the underparts are buff. The female lacks the blue. This species inhabits south-eastern and south-western Australia (predominantly coastal) and Tasmania.

The preferred habitat of this emu-wren is swampy heathlands and low sandplain vegetation. It feeds mainly on or near the ground, normally with the long tail cocked upwards in typical wren fashion. Flights tend to be short and very low, with the long tail streaming behind. The nest is a domed structure with a side entrance, placed near the ground and well hidden in dense vegetation. Two white, red-spotted eggs are laid.

TREECREEPERS

FAMILY *CLIMACTERIDAE*

Although having similar habits of clinging to tree-trunks, limbs and logs, the treecreepers seem not to be related to the sittellas, nor are they related to the northern hemisphere treecreepers or woodpeckers; similarities of feeding behavior have encouraged evolution towards similar shapes and behavior in all groups. These birds climb vertically and often use their stiff tails as props for additional support. The seven Australian species between them cover most of the continent and Tasmania except treeless desert country. The treecreepers are rufous-brown, grey-brown to dark brown birds, with patches and streaks of chestnut, grey, black and white according to species; in flight a broad pale cinnamon patch is conspicuous on each wing. This family is confined to Australia except for one species in New Guinea.

Rufous Treecreeper *Climacteris rufa*

Confined to south-western Australia and extending eastwards as far as Eyre Peninsula in South Australia, the Rufous Treecreeper is an inhabitant of the drier eucalypt forests, woodlands and inland mallee scrublands. In its habits it resembles the other species of Australian treecreepers. These can be distinguished by their characteristic way of spiralling up the trunks and limbs of trees (unlike the sittellas they always move upwards, then fly down to the base of the next tree) pausing frequently to pry an insect from its crevice in the bark. Tree-creepers have a jerking, tail-and-head bobbing progression up the trees. They are also often seen among fallen timber where the decaying wood and loose sheets of bark harbour many insects.

The Rufous Treecreeper nests in a hollow spout or other hole in a tree, sometimes low, occasionally very high; almost invariably a deep cavity is chosen, and warmly lined with bark, fur and feathers. The two or three eggs are white with reddish markings.

WHISTLERS AND THRUSHES

FAMILY *PACHYCEPHALIDAE*

Although not closely resembling the flycatchers, the whistlers and their allies are quite closely related and are sometimes included in the flycatcher family. They are larger and more solidly built birds, slower moving, rarely catching insects in flight but instead finding them on foliage, and especially on the bark of trees. As they have relatively thick heads, they are sometimes known by the common name 'thickhead'. Their bills are strongly developed, with slight hooks at the tip. Although many species are greyish or brownish, some, like the shrike-tits and whistlers, have colourful plumage of yellow or rufous accentuated by black and white. Some species have a powerful and melodious song. The twenty Australian species of this family fall into four genera, the whistlers (*Pachycephala*), the shrike-tits (*Falcunculus*), the thrushes, sometimes known as shrike-thrushes, (*Colluricincla*), and the Crested Bellbird (*Oreoica*).

Rufous Whistler *Pachycephala rufiventris*

The colourful male has a rufous or dull chestnut breast accentuated by a black band across the upper breast and a pure white throat patch. The upper parts are dark olive-grey, and a black band through the eye continues back to meet the black band that separates the white and chestnut on his throat. The female is much less colourful, being buffy streaked with dark brown beneath.

The Rufous Whistler inhabits mostly the drier forests, open woodlands and cleared farmlands, leaving the rainforests and heavy eucalypt forests to the Golden Whistler. It is renowned for its rich and varied song, which includes ringing whip-crack sounds.

This whistler is found throughout Australia in suitable habitats, but is absent from Tasmania. Its nest is a rather frail structure mainly of grass stems and other fine materials, placed in an upright fork of a tree or shrub up to about 10 metres from the ground. It is often possible to see the dark shapes of the eggs through the nest from below.

WOODSWALLOWS

FAMILY *ARTAMIDAE*

Although these birds are somewhat similar to the swallows and martins in their manner of hawking for insects in the air, and in their pointed-winged, forked-tail appearance, the two families are not closely related. They are among the finest fliers of the songbirds (passerines), and among the very few that soar. All species are highly gregarious, normally sleeping, nesting and hunting in parties. When roosting they may pile up together or form close-packed rows along high branches. Woodswallows have long pointed wings which when folded reach almost to the end of the shallow-forked or blunt tails. They have short, pointed conical bills with small bristles at the base. Their plumages are mostly brownish or slaty grey above, white below, with chestnut underparts on one species. Woodswallows are unique among songbirds in having powder-down feathers, the tips of which disintegrate to form a talcum-like powder used in dressing the feathers. Six of the ten species occur in Australia, and four are endemic; it is thought that this family may have evolved within Australia, and later extended its distribution northwards.

Masked Woodswallow
Artamus personatus

An inhabitant of the inland savannah and open woodlands almost throughout Australia, but absent from Cape York, the heavier-forested areas of the south-east and south-west, and Tasmania.

The face and throat are black, which surrounds the eye and extends forward to the base of the bill, forming a conspicuous mask. The plumage of the upper parts is a light dove grey, mostly separated from the black mark by a white line. The undersurfaces are greyish-white, and the tail is tipped with white. The female has less contrast in its plumage pattern, being dusky on the face and throat, blending more gradually into the greys of the body.

The Masked Woodswallow is a highly nomadic species, wandering over great distances in flocks searching for the most favourable conditions. When a favourable area is found — usually a region that recently experienced substantial rains and will in the following few months have lush growth and abundant insect life — the birds remain to breed, irrespective of the time of year. The nest is a loosely constructed cup-shaped structure usually in a bush not far above the ground; the two or three eggs are variable in colour, pale brown to greenish, with grey and brown markings.

TYPICAL OWLS

FAMILY *STRIGIDAE*

The owls of this family are more hawk-like in appearance than the barn owls; the feather discs around the eyes are smaller, and remain two separate circles instead of joining to form a single large mask. The body plumage is, in Australian species, boldly barred or streaked. Like the barn owls, these have very strong legs with powerful curved claws. Both the eyes and ears are directed forward, and cannot be moved in their sockets, so that the bird must turn its head to follow movements. The bill is hawk-like, sharply hooked. The typical flat conformation of the owl's face is built up of outwards-radiating feathers which are of a peculiar hard wiry texture, and form a disc (resembling a radar tracking antenna) which probably serves some acoustic function, perhaps giving greater sensitivity and directional accuracy to these birds' hearing. Four of the total 120 species occur in Australia, one being endemic.

Boobook Owl *Ninox boobook*

This small owl is brown on the upper parts, marked with a buff and spotted with white on the wing coverts. The underparts are brown, longitudinally streaked with buff. The area around the eye is very dark, and there is a lighter patch on forehead and chin.

The Boobook Owl is distributed throughout Australia in forests, open woodlands, and in desert country wherever there are trees (as along watercourses) with hollows large enough for roosting and nesting. The two or three white eggs are laid on the wood dust at the bottom of a hollow. The Boobook preys upon small animals and insects. Its call is a double-noted 'boo-book' or 'mo-poke', often mistakenly attributed to the Tawny Frogmouth. There are other calls less commonly uttered. When feeding young in the nest a Boobook family carries on quite a 'conversation', maintaining contact perhaps, with a variety of calls rarely if ever heard at other times.

Mammals

It is well known that Australia has become a fauna region distinct from the rest of the world. It has long been cut off by oceans from all other land masses, and this has isolated its creatures from the evolutionary changes and upheavals that occurred elsewhere on the globe. But it seems to have been the Australian mammals more than any other group which have been most influenced by the ages of isolation. The birds, while including many unique elements, have understandably maintained greater contact with the bird populations of other lands. There are for example the constant comings and goings of the many migratory birds, and even small land birds are occasionally blown in storms across narrow straits. The reptiles too seem to have maintained more closely their links with other continents. But only in comparatively recent times have placental mammals, the bats, flying foxes, rodents, the Dingo and Aboriginal man, found their way to Australia.

Among the mammals, Australia's marsupials and monotremes are the most distinct groups; these kinds of mammals have long been extinct on most other continents. Early primitive mammals were marooned on this ancient continent and found it a sanctuary isolated from competition from the more advanced kinds of mammals which were evolving on other continents.

The theory of continental drift probably best explains Australia's long period of total isolation and subsequent renewal of contact with the rest of the world. Australia appears to have been part of a larger land mass some forty-three million years ago. This huge continent broke apart, and the slabs or plates of the earth's crust carrying the smaller segments drifted gradually apart to become South America, Antarctica and Australia. All have closely related plants and animals, living, or as fossils in rocks laid down at that time.

Australia drifted for thirty million years or more, completely isolated from any other land mass. During this time its animal life was cut off from outside influences. The marsupials and monotremes evolved towards their present multiplicity and diversity of forms, and probably made up the entire mammal fauna.

Eventually, some ten million years ago, this huge drifting plate of the earth's crust carrying Australia and New Guinea approached what are now the islands of Indonesia, the edge of the Asian land mass. At this time the mammals of Asia began to find their way into Australia. At first came those that could cross the gradually narrowing stretch of ocean — the bats and flying foxes, the rodents perhaps riding on driftwood blown south by cyclones. Finally came Aboriginal man, bringing a type of domesticated dog which, running wild and spreading during the following thousands of years, became the distinctive reddish-yellow Dingo.

Within Australia the mammals, whether monotremes, marsupials or placentals, have encountered a great diversity of environmental conditions. In spreading across the face of the continent to inhabit these vastly differing regions, the first marsupials, and later the placental mammals, became in time separated from others of their own kind, each isolated population gradually aquiring characteristics to suit the local habitat, and eventually becoming different enough to be each a distinct new species.

Over the millions of years the mammals have diversified to fill almost every available habitat from jungle canopy to desert dunes until, by the time of arrival of European man, there were about 120 species of marsupials, 108 species of native placental mammals, and two species of monotremes in Australia.

Australia and New Guinea are the only places on earth where the three basic types of mammals exist side by side; the greatest, most fundamental differences between them are in their ways of reproduction. With marsupials, the young are born at an age and size which seems incredibly premature, because there is no provision within the womb for nourishment of the foetus once it has grown beyond this size. Immediately after birth the tiny young one struggles unaided through the fur to the pouch where it becomes firmly attached and remains for a long time.

The placental mammals, however, have succeeded in perfecting a device that enables the foetus to remain in the perfect environment of the womb for the full term. This structure is the placenta, which brings bloodstreams of mother and developing young into contact for exchange of the elements essential for respiration and growth.

The young of the largest marsupial, the Red Kangaroo, is only three-quarters of an inch long at birth, but some young placental mammals (such as foals and fawns) can run with their mother soon after birth. The superiority of the placental mammals lies in the shorter total time required, enabling them to multiply faster than most comparable marsupials.

The third major group of mammals are the monotremes, of which there are only two Australian species, and several in New Guinea. These animals are considered to be relics, of ancient origin, very primitive in their similarity to reptiles, especially in their egg-laying way of reproduction. They are, however, extremely specialized — the Platypus for its amphibious way of life, and the various spiny anteaters for their ant diet and protective spines — and it is probably because of their success in this specialization that they have never been pushed from their niches by any marsupial equivalent.

Australia's marsupials can be considered in four major groupings — the marsupial carnivores, the kangaroos and wallabies, the possums wombats and Koala, and the bandicoots. Each of these groups has evolved to occupy a place in the Australian environment which on other continents is the role of an entirely different animal. Many of these Australian animals, although of another origin, have over the millions of years come to resemble animals which follow a similar way of life in another part of the world. This tendancy for animals of different origins to acquire similar characteristics is known as convergent evolution.

Examples of this are not difficult to find. The Tasmanian Wolf or Thylacine has an appearance remarkably like the unrelated wolf of the northern hemisphere; each has acquired this shape because it is apparently the best mammal pattern for an animal following that particular way of life. In our carnivorous tree-climbing marsupials we see such a resemblance to the unrelated feline cats that we call them native-cats. Both marsupial and placental cats have evolved a similar shape because this is the superior form for an arboreal, nocturnal, predatory mammal to take.

But animals of similar habits do not always aquire such similarity of appearance and actions. Australia's large herbivorous mammals, the kangaroos and wallabies, are totally unlike the grazing animals, the deer, antelope, horses and cattle of other parts of the world. For some reason the large Australian herbivores have chosen the hopping rather than the running way of fast travel. But even in this instance, convergent evolution has produced remarkable similarities of head, jaws and teeth, as all are grazing animals.

In Australia, possums, and cuscuses take the place of monkeys, and gliding phalangers such as the Sugar Glider resemble the gliding squirrels of other lands. The Platypus is rather like the beavers, sharing with them the flattened paddle-like tail and other adaptations for swimming.

Australia's placental mammals include animals as large as the Sea Lion and as small as the native-mice. Some, like the hopping-mice, have adapted so well to the deserts that they can live without any water under hot dry conditions. Others, the water-rats, have joined the Platypus in rivers and creeks, and have webbed feet. Some of the Australian native-mice and native-rats have been so long isolated on this continent that they are referred to as the 'old endemics'. In this group are the hopping-mice, tree-rats, stick-nest rats, rock rats and the many little native-mice.

BANDICOOTS

FAMILY *PERAMELIDAE*

Bandicoots on the whole are mixed feeders, that is, they eat insects, vegetable matter of many kinds, occasional very small mammals and lizards. The food preferences vary considerably from one genera to another, some groups being entirely insectivorous, others vegetarian, and some accepting a mixed diet. These varied feeding habits are reflected in certain main characteristics by which bandicoots are classified: they possess the many-incisored dentition typical of the marsupial carnivores, but have the hind-foot combined toe structure of the herbivorous kangaroos.

There are two main evolutionary lines of bandicoots. One comprises the coarse, almost spiny-furred ordinary bandicoot and the Pig-footed Bandicoot; the other is represented by the softly silken-furred, long-eared rabbit bandicoots. A number of species dig burrows, while others nest under dense vegetation or among logs or boulders.

The pouch opens in a rearward direction on all bandicoots. Some members of this family have become very rare, possibly extinct, while many have become uncommon and even those that were abundant have suffered great loss of numbers. They are useful little animals, even when they dig in gardens, for they consume a great quantity of harmful insect larvae. There are five genera — the short-nosed bandicoots (genus *Isoodon*), the long nosed (*Perameles* and *Echymipera*) rabbit-eared bandicoots (*Macrotis*), and the Pig-footed Bandicoot (*Chaerops*).

Long-nosed Bandicoot *Perameles nasuta*

The long-nosed bandicoots, of which there are five species (in two genera, *Perameles* and *Echymipera)* are more lightly and gracefully built than the common short-nosed bandicoots, and their snouts are extremely elongated, tapering to a slender point. Their ears are longer (though still much shorter than those of the rabbit-eared bandicoots) and pointed, and the fur less coarse than the short-nosed bandicoots. The long-nosed bandicoots are nocturnal and carnivorous, feeding principally on small invertebrates such as earth worms, spiders, insect larvae and the like.

The common Long-nosed Bandicoot inhabits rainforests, wet and dry eucalypt forests, and woodlands of eastern Queensland, eastern New South Wales and eastern Victoria. It is common in the Sydney area where its scratching about in search of beetles and other insects may at times damage surburban lawns. In this district breed-ing takes place throughout the year, and the young are born after a gestation period of about two weeks. Two or three make a normal litter, but there may be up to five; the young leave the pouch at the age of nine or ten weeks. Long-nosed Bandicoots are nocturnal and solitary in habits — mating is the only occasion that this pattern briefly changes.

Pig-footed Bandicoot *Chaeropus ecaudatus*

Looking like a minature deer with its long ears, long slender legs and slightly crested tail, this rabbit-sized bandicoot is one of Australia's rarest animals. In fact is is probably extinct, the last specimen having been caught in 1907; the last reports of sightings, in 1926, came from the aborigines of the Musgrave Ranges in the desert heart of Australia.

The most distinctive feature of this bandicoot, the only species of the genus, is the pig-hoof like structure of its feet. On each hind foot is one large functional toe, and on the forefeet only two functional toes which together resemble the cloven hoof of a pig. In the days of earliest settlement of Australia this dainty little bandicoot inhabited large areas of the arid and semi-arid woodlands, mallee scrub and grasslands of Central Australia, inland Western Australia, South Australia except the south-east, north-western Victoria and south-western New South Wales.

Not surprisingly there is very little recorded information on the natural history of this species. The breeding season appears to have been in winter, and although the pouch contained eight teats, it seems that no more than two young were ever found in a pouch. Early observations suggest that they were nocturnal, and probably purely vegetarian; they spent the day in a nest of sticks and leaves, which may be in a shallow hole in the ground. The disappearance of this unusual, attractive and harmless little marsupial was due to the trampling and destruction of its natural habitat by flocks of sheep and herds of cattle, which left it increasingly vulnerable to predation by introduced foxes and cats, both of which now occur wild in the most remote parts of Australia.

Rabbit-eared Bandicoot *Macrotis lagotis*

Among the most graceful and beautiful of all native mammals, this bandicoot has large rabbit-like ears and very long silky black and white tail. It is about forty-five centimetres in length, plus a further twenty-five centimetres for the tail, and

was also known as the Dalgyte or Bilby. When running this animal moves in a graceful flowing canter on quite long legs, the silken tail held aloft like a banner.

Unlike other bandicoots, the Rabbit-eared always lives in a burrow which may be from one to two metres in depth and descends spirally. Within these burrows they are able to escape the daytime heat of the desert regions which constitute the major part of their habitat. They are accomplished diggers, each paw being armed with strong claws, those of the fore-feet scratching out the soil, those of the hind feet kicking the loose material backwards. The burrows are blocked at intervals by plugs of soft earth.

The common Rabbit-eared Bandicoot once had a wide distribution in Western Australia from the Kimberleys to the south-west, in western and northern South Australia, western New South Wales, and far south-western Queensland. It is now found only in remote arid regions. A second species, the Lesser Rabbit-eared Bandicoot or Yallara, which has an all-white tail, inhabits the sandhills of the Simpson Desert region of central Australia.

Sugar Glider *Petaurus breviceps*
One of the most abundant of Australia's 'flying' marsupials, the squirrel-sized Sugar Glider has successfully occupied a wide range of habitats from tropical woodlands to alpine forests. It is a fast-moving nocturnal hunter of insects and nectar, and travels through the bushland by gliding from tree to tree.

Tiger Cat *Dasyurus maculatus*
Equipped with jaws that open tremendously wide, the Tiger Cat is a powerful predator of Australian forests, hunting mostly in the treetops at night, and preying upon birds and small mammals. It is the largest marsupial carnivore on the Australian mainland, having a total length of about one and a quarter metres.

Dibbler *Antechinus apicalis*
One of the rarest of all Australian marsupials, the Dibbler or Freckled Marsupial-mouse is a fierce hunter of insects and other small creatures. Since 1884 only three of these shy nocturnal creatures, which have a total length of about twenty-five centimetres, have been caught, and it is possible that this species is on the verge of extinction, if not already gone for all time.

Quenda *Isoodon obesulus*

The bandicoots of the genus *Isoodon* are known as short-nosed bandicoots; although all marsupial bandicoots have long slender snouts, the members of this genus are short-nosed by comparison with some other extremely long-nosed genera.

The Quenda, also known as the Brown Bandicoot or Southern Short-nosed Bandicoot, is about the size of a rabbit, with a head and body length of approximately five centimetres and a short tail of twelve centimetres. The muzzle is pointed, the ears are short and rounded, fur coarse and grizzled yellowish-brown, white on

the belly. This species is distributed down the east coast from Cape York Peninsula through eastern coastal New South Wales to southern Victoria and south-eastern South Australia, and occurs in Tasmania and south-western Australia.

Quendas are nocturnal and insectivorous, inhabiting forests and heathlands wherever there is dense ground cover. Their presence can be detected by the little conical pits left where they have been digging for insect larvae and beetles.

A rough nest of sticks is constructed on the ground in dense vegetation. There is usually no entrance as the bandicoot just burrows under the pile of debris. Seven young form the usual complete litter.

Fat-tailed Dunnart *Sminthopsis crassicaudata*
The mouse-sized Dunnart is a fierce little nocturnal predator which hunts for grasshoppers, spiders and other small creatures on the ground. These it will attack with all the ferocity of a miniature native-cat, although it is only twelve centimetres in length, including the fattened tail which serves as a reserve for times when food is scarce.

Numbat *Myrmecobius fasciatus*
So distinctly banded that it cannot be mistaken for any other creature, the Numbat uses its strong claws to scratch into the tunnels of termites. With an amazing flicking of the long thin extensile tongue, which shoots out from the snout at all angles to follow the intricate passageways of a termite nest or the galleries of termite-ridden timber, the marsupial anteater feeds upon 'white ants' by the thousand. In marked contrast to the nocturnal life of most marsupials, this beautiful creature is out of its hollow log home and feeding during the day.

Western Native-cat *Dasyurus geoffroii*
Although cat-like in size, in their secretive nocturnal habits, tree-climbing skill, stealth in hunting and effectiveness at the kill, the Australian native-cats are creatures of totally different origins. As marsupials they are much more closely related to kangaroos and koalas than to ordinary cats; their likeness to other cats results from the similar predatory way of life. This Western Native-cat is one of three very similar species.

CARNIVOROUS MARSUPIALS

FAMILY *DASYURIDAE*

The carnivorous marsupials are all predators. The larger hunt wallabies, birds and reptiles, while the small species, by far the greater part of the family, are insectivorous. All carnivorous marsupials have three pairs of lower jaw incisors, five toes on the front feet, and never less than four on the hind feet. Together these features distinguish them from all other marsupials. Although given such names as 'cat, wolf, tiger', they are quite unrelated to any of those animals, all being typical marsupials in that they give birth to incredibly tiny young which must complete their growth, up to the stage at which other mammals are born, in a pouch. However the pouch of many marsupial carnivores is not the deep pocket-like pouch of the kangaroos, but a shallow depression where the tiny young attach to a nipple and cling among long wiry fur.

Many of the small insectivorous marsupials such as the Dibble are easily mistaken for rats or mice, and are often known as 'native-mice', but any similarity to ordinary mice is superficial — they are more like tiny cats, equally bold and ferocious when hunting. The alert little marsupial-mice have sharp-pointed faces and needle-like teeth. Marsupial carnivores occur not only in Australia, but also in South America, and one species, the opossum, in North America. They are probably the most like the original ancestral marsupials which first reached Australia, from which the present great variety evolved. There are thirty-eight species.

Brush-tailed Phascogale
Phascogale tapoatafa

There are two species of the genus *Phascogale*, both brushy-tailed, carnivorous, rat-sized mar

supials, known also as 'wambengers' or 'tuans'. The Red-tailed Wambenger, *Phascogale calura*, has the part of the tail nearest the body bright rufous; the Brush-tailed Phascogale or Tuan has an entirely black long bushy tail, which often has its wiry fur bristling out like a bottlebrush, particularly when the animal is very active or excited. The Tuan is almost exclusively a tree dweller, climbing with great agility, descending treetrunks head first with body pressed against the bark, much in the manner of a squirrel. It preys upon insects, small lizards, small birds roosting in the treetops at night, and probably partly responsible for the many birds nests that are robbed. The Brush-tailed Phascogale is about

forty-five centimetres in length, of which almost half is tail. It inhabits rainforests, eucalypt forests, woodlands and wooded grasslands of much of northern, eastern, south-eastern, south-western and parts of the interior of Australia. A hollow of a tree is its usual home. Three to six young are the usual litter, and these remain attached to the pouch area for about four months.

Mulgara *Dasycercus cristicauda*

This inhabitant of the deserts of the northern half of Australia appears superficially to be a terrestrial equivalent of the phascogales; in actual fact very little is known of its true relationships to other marsupial carnivores. It is approximately the size of a large rat, but has the attractive alert, pointed-nosed, big-eyed face typical of the insectivorous marsupials. Its tail is quite long, thick and with a crest of shining black hair along the upper part towards its end. The remainder of the fur is a contrasting reddish or sandy brown.

The Mulgara is said to be one of the most fearless and intelligent of the smaller marsupials, and a most efficient predator, seizing mice with a quick grab at the nape of the neck, and killing them instantly. Although mice, birds and small lizards are always killed in a lightning rush, some beetles and other insects are first picked up cautiously in the hands. These active animals consume up to twenty-five per cent of their own weight in meat each day. They have physiological adaptations which enable them to live in some of the most arid regions of Australia without drinking water or even eating succulent plants. The kidneys are extremely efficient and excreted urea is very concentrated, minimizing water loss. The Mulgaras avoid the heat of day by staying in their burrows.

Red-tailed Phascogale *Phascogale calura*

Also known as the Red-tailed Wambenger, this species is smaller than the Brush-tailed Phascogale; its fur is brown with the base of the tail between body and brush bright rufous. It inhabits inland parts of the southern half of Western Australia, the southern part of the Northern Territory, north-western Victoria and south-western New South Wales. It is probably rare except in the more remote areas. This species also is almost exclusively arboreal and nests in hollows of trees.

The name 'native-rat' sometimes applied to these and other marsupial carnivores of similar

size is not very appropriate so they are not, in appearance, even superficially rat-like. Both the Red-tailed and Brush-tailed Phascogales are distinct from all other of the broad-footed marsupial-mice (i.e. the arboreal small marsupial carnivores) by the unique tails with their silky brush of long black hairs, which may either stand out or lie sleekly flat. These two species are also the largest of the broad-footed group.

Tasmanian Devil *Sarcophilus harrisii*

Among the marsupial carnivores, the Devil is exceeded in size only by the Thylacine. It is entirely black except for a few small white markings, and its eerie whining snarl must be at least partly responsible for the name 'devil'. About the size of a small dog, this predator has a short, heavy-set body, short powerful legs, a disproportionately large head, and short tail. This is one of the few marsupials that does not have an elongated muzzle; its head is rounded, broad and heavily built. The upper jaw projects over the lower, and both the head and jaws are extremely powerfully muscled. The jaws and teeth, like those of the hyaena, are adapted for crushing bone as well as for shearing and tearing flesh.

The Devil runs in a peculiar stiff-bodied canter, almost continuously nosing the ground. Almost anything it can kill, together with carrion, makes up the diet. Tasmanian Devils are mainly nocturnal, and in their prowlings often fight noisly among themselves. Although they do not

burrow, their claws are long, and they can manage to scramble up trees with a sloping trunk.

The Tasmanian Devil's lair may be a cave or a crevice beneath boulders, or under the roots of a fallen tree. A bed of bark, leaves and grass is constructed. Four young are born, quickly disappear into their mother's pouch, and do not emerge for some fifteen weeks. These marsupial carnivores once inhabited the mainland, but appear to have been replaced there by the dingo; they are still quite common in Tasmania.

Thylacine *Thylacinus cynocephalus*

The largest of Australian marsupial carnivores, the Thylacine, perhaps better known as the 'Tasmanian Tiger' or 'Tasmanian Wolf', was persecuted by the farming community until extinct, or very nearly so, for its tendency to add an occasional sheep to its usual diet of small wallabies and other native game. It is believed that a few may have survived in remote parts of mountainous and heavily forested Tasmania, and that it may be gradually increasing in numbers. The Thylacine was about the size of a moderately large dog, with a heavy kangaroo-like tail merging gradually into the hindquarters. Its fur was olive-brown with darker transverse stripes across the upper parts from shoulders to tail, and the massive, very dog-like head had exceptionally wide-opening jaws. Although this animal occurred only on the island of Tasmania at the time of European settlement, fossil evidence shows that it once occurred throughout most of Australia and in the highlands of New Guinea. A mummified specimen found in a cave on the desert-like Nullarbor region of Western Australia still had fur with clearly visible stripes on the mummified body, and was estimated to have died no more than five thousand years ago.

Tiger Cat *Dasyurus maculatus*

The Tiger Cat, although hardly the equal of a true tiger, is the largest arboreal marsupial carnivore, growing to a length, nose to tail-tip, of about 1.2 metres. Like the Thylacine, the Tiger Cat has jaws that open exceptionally wide; it has a reputation, dating back to the earliest days of settlement, for ferocity and stubbornness. It is principally an animal of the treetops, occurring in greatest numbers in the heavily forested coastal mountainous regions of eastern and south-eastern Australia and Tasmania.

The tree-climbing skill of the Tiger Cat is largely due to the special adaptations of the hind feet, which have ridged rubbery pads on the soles, and sharp claws. Birds roosting at night probably make up a large part of its prey, but it is also able to kill very small wallabies, and at other times probably finds subsistence from reptiles and large insects.

There are many tales of the boldness of the Tiger Cat. The presence of one of these car-

nivores is usually first noticed when there are nocturnal killings of roosting fowls. If trapped or cornered the Tiger Cat, with its widely opening jaws and canine teeth is not to be despatched easily. One was reported to have kept two terriers at bay from its refuge in a hollow log; another killed a large tomcat in a fight. The Tiger Cat has become rare in many parts of Australia. Four to six young are born about May, and carried in a rudimentary pouch, a crescent-shaped flap enclosing the front and sides of the mammary area.

Yellow-footed Marsupial-mouse
Antechinus flavipes

Measuring just 20 centimetres from pointed nose to tail tip, the Yellow-footed Marsupial-mouse is a fierce little hunter, with a voracious appetite for beetles, grasshoppers, spiders and small lizards. Like other marsupial-mice this species has an armoury of sharp-pointed teeth. Its head and foreparts are dark grizzled grey, the lower back and rump brown or with reddish tinge, and its flanks are reddish brown. The upper surfaces of its feet are creamy white to brownish yellow, and there is a yellowish to brownish eye ring; the tail is hairy, but tapering rather than crested or bushy. The undersides of its broad feet have ridged pads which enable it to climb on rocks and bark, in contrast to the purely terrestrial species of marsupial mice which have narrow smooth-soled feet. The Yellow-footed Marsupial-mouse is believed to be typical of the most primitive and generalized of all Australian marsupials, similar to the earliest types from which the present great variety has evolved. It lives in rock crevices, logs and hollow limbs of trees, and inhabits rainforests, eucalypt forests and woodlands of eastern and south-western Australia.

Dibbler *Antechinus apicalis*

This carnivorous marsupial is one of Australia's rarest mammals. Until January 1967 it had not been seen for eighty-three years, and consequently was considered possibly extinct. It is thought by some zoologists to be a connecting link between the insectivorous marsupial mice and the larger carnivorous native-cats.

The Dibbler, known also as the Freckled Marsupial-mouse for the white speckling of its grey-brown fur, is about twenty to twenty-five centimetres in total length. Its most striking feature is a clear white ring of fur around each eye, visible from a distance of several metres. Distinctive also is the tail, which tapers strongly but uniformly from a wide base to a fine-pointed tip, due to the fur at the base of the tail being long, and becoming progressively shorter towards the end of the tail. The speckled effect in the fur is due to each long hair being tipped with white. The fur of the forearms is distinctly reddish.

The distribution of this marsupial at time of European settlement extended over quite a large area of the south-western corner of Western Australia, from the Moore River on the west coast, to the south coast east of Albany. It is now known to occur only on the Mt. Manypeaks range, east from Albany, where a small colony was discovered.

Since 1967 further attempts to find living specimens of the Dibbler have been unsuccessful, even at the site of the 1967 rediscovery. The Dibbler remains not only one of the rarest marsupials (only three being caught in eighty eight years). But in the last seven years, that small colony, perhaps the very last, may have gone.

Dunnart *Sminthopsis crassicaudata*

Although generally known simply as the 'Dunnart', the full name of this species is 'Fat-tailed Dunnart', for there are seven other less well known species which have become rare or inhabit remote areas. The Fat-tailed Dunnart is a mouse-sized carnivore with big ears, a sharp pointed face, bulging bright black eyes, and a long tail that is fattened in the central part and narrows where it attaches to the body. Like other dunnarts, it is an active hunter, attacking grasshoppers, spiders and even mice on the ground at night. Its teeth, unlike those of ordinary mice, are needle-pointed, cat-like.

The Fat-tailed Dunnart inhabits eastern Australia from south-eastern Queensland, western New South Wales, western Victoria, eastern South Australia, to the south-west of Western Australia.

The fat tail appears to be a food storage, necessary because all fat-tailed species inhabit inland areas varying from moderately dry to extremely arid. The tail becomes fattened and spindle-shaped in good seasons when food is abundant, but very thin in times of drought. The Fat-tailed Dunnart nests in hollow logs and fence posts, or under stones, stumps or clumps of dense vegetation. In cold weather and times of food shortage the Dunnart is able to become torpid, a state of sluggish semi-hibernation, thereby reducing energy requirements and living on stored food.

Numbat *Myrmecobius fasciatus*

The Numbat or Marsupial Anteater is an aberrant member of the native-cat family, Dasyuridae, but is in many ways so different that it is sometimes placed in a separate family of its own. One of the most beautifully patterned of all Australian mammals, the Numbat is in general colour reddish-brown to grey-brown, with prominent white and black stripes across back and rump. The face is sharply pointed, and the long slender tongue can be extended ten centimetres beyond the nose tip. Its teeth are extremely small and degenerate, and vary in number from fifty to fifty-two, the largest number found in any land mammal. The Numbat is about the size of a rabbit, twenty-three to thirty centimetres for the head and body, and another seventeen to twenty for the bushy tail.

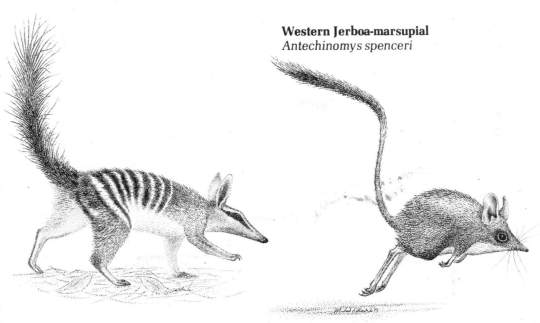

The Numbat is very closely tied to the natural Wandoo woodland country, which is the principal habitat of the surviving population in south-western Australia; the species has become extinct in eastern Australia. This woodland is of scattered trees above patches of low scrub and bare ground. Most of the mature Wandoo trees have been eaten hollow by termites and the ground throughout the area is littered with limbs and logs that almost invariably are hollow. These hollows provide this defenceless creature's only protection against predators (foxes and eagles) while the termites themselves are its main food.

Unlike the majority of small mammals, Numbats are abroad in full daylight. They are solitary animals which spend most of their time scratching in the ground and turning over pieces of wood in their search for termites. If disturbed they run very swiftly to the nearest hollow log.

The young are born January to April. As the female Numbat has no pouch the four young (the normal litter size) cling in the long hairs of the pouch area under the mother's belly, but are held even more securely by the teats which, as with all marsupials, swell in their mouths to make a connection that will not be broken for many months. At this stage while they are less than twenty or thirty millimetres in length, the female seems unaware of their presence. Later, when no longer attached to her, they are left in a hollow log while she is foraging during the day.

Western Jerboa-marsupial
Antechinomys spenceri

Known also as the Wuhl-wuhl or Pitchi-pitchi, this tiny kangaroo-like marsupial has a resemblance to the rodent hopping-mice, although they are not related. The Jerboa-marsupial is a mouse-sized insect hunter of the desert grasslands, spinifex, saltbush and open woodland country of the interior. The distribution of this species extends right through central Australia from western Queensland to the interior of Western Australia. A second species, the Eastern Jerboa-marsupial or Kultarr, inhabits the inland south-eastern regions.

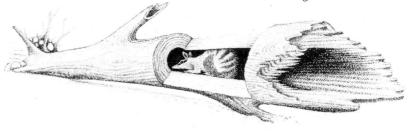

The Jerboa-marsupial has extremely long legs, particularly the hind legs, and a long brush-tipped tail. It was once thought that, because of their kangaroo-like shape, they must leap along, but recent studies by high-speed photography and foot-print patterns have shown that their movement is more like that of a sprinting greyhound, although the forepaws touch the ground only lightly just ahead of the hind paws, which give the powerful forward thrust.

Western Native-cat   *Dasyurus geoffroii*

At first glance this marsupial carnivore appears to be a smaller edition of the Tiger Cat, but differs in having a sharper-pointed face, ears relatively larger and more pointed; the tail is not spotted like the rest of the body, and the soft pads on the soles of its feet are granular-surfaced instead of ridged.

The Western Native-cat was once very widespread, occurring in western New South Wales, in Queensland except for the northern half and the south-east, in north-western Victoria, South Australia except the south-east, and through Central Australia into the southern half of Western Australia. This species is now probably extinct in eastern Australia, but remains quite common in the forested south-western corner of Western Australia, where there occurs a larger sub-species.

Like the two other species of native-cats, the Eastern Native-cat or Quoll, and the Little Northern Native-cat, this animal is a nocturnal carnivore of the forests and woodlands. It hunts in the trees and on the ground. The native-cats are of zoological interest in presenting one of the few cases of over-production of young. As many as eighteen may be born at the one time, but only eight can be reared. The first (probably the strongest and best suited to perpetuate the species) are the only ones to be able to attach to a teat, an attachment which cannot be broken for several months.

Native-cats on the whole are very beneficial, destroying a large number of pest creatures including rabbits, mice and the larger insects.

CUSCUSES AND LARGE POSSUMS

FAMILY   *PHALANGERIDAE*

The members of this family are all tree-dwellers which feed upon foliage, flowers and native fruits. Their tails are long and well adapted for tree climbing, being prehensile, with bare rough skin along the undersurface and thus able to grip very securely when wrapped around a branch. The tails are often curled into a tight roll when not being used for climbing. The feet have large curved claws, except for the 'big toe', which is, like a human thumb, opposable to the other four toes and permits a secure and reliable grip of branches. It is a characteristic of all these, and also the smaller possums and gliders of other families, that the small second and third toes of the hind feet are joined together up to the outermost joint. The twin claws of this double toe are used like a comb for the fur. All species have a deep pouch which opens forwards.

The Cuscuses are distinguished by having the outer half of the tail completely naked of fur and covered with rough scales of rasp-like pattern. The ears of these large animals are scarcely visible being short and set in thick fur, and this gives them a most distinctive appearance. The large possums include the brushtail possums, and the strange and uncommon Scaly-tailed Possum. The brushtail possums have a thick, bushy tail with only a small naked area beneath the tip, long ears, and pointed, rather fox-like faces. The Scaly-tail is in some ways intermediate between brushtails and cuscuses; its tail entirely furless and covered with rough scales.

Common Brushtail Possum
Trichosurus vulpeca

A large and solidly built possum, having a head-and-body length of around forty-five centimetres, plus a further thirty centimetres for the bushy tail. The body fur is usually silvery-grey, but sometimes almost black, light yellowish-grey on the undersurfaces; the tail is darker black towards the end, except in Western Australia where the tip is usually white.

This possum, known in different districts as the Common Possum, Brush Possum or Brushtail Possum, is of very widespread occurrence, and there are many slightly different geographical races. It occurs in Tasmania, south-eastern, eastern, northern, south-western and many parts of inland Australia. Although the word 'possum' is derived from 'opossum' the name given to a North American marsupial, the Australian animals are quite distinct, to the extent that the shortened name has not only long been correct, but also serves to emphasize the difference.

The Common Brushtail is indeed common, for it is one of our most successful marsupials, and has maintained or increased its numbers in spite of man's changes to the environment. It shows its adaptability by frequently taking up residence in house roofs instead of tree hollows. Where trees are scarce it lives in rock crevices or rabbit burrows. Among the approximately eleven subspecies are the Coppery Brushtail of North Queensland, South-western Brushtail, and the Tasmanian Brushtail, each well defined in fur colours, markings and distribution.

Scaly-tailed Possum
Wyulda squamicaudata

The unique Scaly-tailed Possum is at once distinguishable from all others by its entirely furless tail which is covered for its full length on all sides by rasp-like scales.

This is one of the least-known of the larger Australian possums, and possibly one of the most uncommon. As it is confined to one of the most remote and inaccessible regions of Australia — the north-western parts of the Kimberley district of Western Australia — it was not discovered until 1917.

Not until 1954 was another found, this time from a remote part of the Kimberley coast. In recent years it has been discovered that the possum is not so extremely rare as at first appeared. The scarcity of specimens was probably because it lived where European man had but rarely penetrated.

The Scaly-tail feeds largely upon blossoms, and appears to spend much time among the tumbled boulders of extremely rugged sandstone country. All that is known of its breeding is that the single young may be found in the pouch between June and March.

Spotted Cuscus   *Phalanger maculatus*

These very monkey-like phalangers are large — head and body length being around seventy centimetres, and the tail a further forty-five centimetres. Their fur colour is variable, grey to rusty brown, spotted or plain, or sometimes entirely white. The Spotted Cuscus inhabits the rainforests and gallery forests of north-eastern Cape York, in northern Queensland; their distribution extends southwards about as far as the McIlwraith Range. A second species, the Grey Cuscus *(Phalanger orientalis),* which resembles the Spotted but has a dark dorsal stripe from the top of the head to the rump, inhabits rainforests of the north-eastern corner of Cape York. Both occur also in New Guinea, the true home of this genus of marsupials.

The slow-moving nocturnal cuscuses are entirely arboreal, and live upon leaves and wild fruits. They spend the daylight hours curled up in the high forks of trees among thick foliage, and when they do move, seem very sluggish with their slow deliberate movements. During the night they are much more active, consuming large quantities of leaves. The pouch may contain from one to four young.

DINGO

FAMILY   *CANIDAE*

Australia has but one species in this family, and this is an animal so recently arrived from the northern continents (probably coming with the aborigines as a domesticated dog) that it barely qualifies as a native animal. It is in fact a well established introduced animal, a sub-species of the ordinary dog.

This sole representation of the family has, however, now become an established part of the Australian fauna (as have the rabbit, fox and cat) and over the thousands of years since its arrival has had a very considerable impact upon the older fauna, particularly the indigenous marsupial carnivores. Competition from the more advanced, more intelligent Dingo was probably the reason for the extinction on the Australian mainland, a few thousand years ago, of the Thylacine and the Tasmanian Devil, both of which remained plentiful in Tasmania, which has no dingos.

Dingo   *Canis familiaris*

Australia's wild dog is not easily distinguished from many ordinary dogs of similar size except by a number of small features combined — the ears remain always erect the tail is bushy, the canine teeth average somewhat larger, it utters yelps and howls instead of barking, and there are differences in basic patterns of behaviour.

Colour is variable, most commonly tawny yellow with a paler belly, white tail tip and feet.

The Dingo occurs throughout Australia, but has been driven back from the more closely settled areas; it is absent from Tasmania.

The close relationship between the Dingo and ordinary domesticated dogs is shown by their ability to interbreed, but this probably occurs far less frequently in the wild than commonly supposed. Variations in fur colour, often said to prove crossings with ordinary dogs, is a natural feature of the Dingo. The earliest explorers, pushing far into regions never before seen by white man, recorded sightings of Dingos of a variety of colours.

FALSE-VAMPIRE BATS

FAMILY *MEGADERMATIDAE*

The members of this family of large bats are characterized by their long narrow erect nose-leaves, and large ears, which join together on top of the forehead for at least half way up their inner edges and have a divided fleshy projection or tragus in each ear opening. There are no upper incisor teeth and the tail is so short as to be unrecognizable externally. There are five genera in the family, only one of which occurs in Australia, and this genus *(Macroderma)* is represented by a single species. Other members of the family occur in Asia and Africa. The genus occurring in Australia is one of the most interesting zoologically. The one Australian species, the Ghost Bat, is not only the largest of the leaf-nosed bats, but also the largest of all the insectivorous group (and has therefore been able to extend its food source to larger creatures). It is not a vampire or blood-sucking bat, but to a degree is cannibalistic, taking many smaller bats.

Ghost Bat *Macroderma gigas*

This very large bat, which has a wingspan of about sixty centimetres, is a highly specialized carnivore, preying upon small mammals, lizards, roosting birds and even other smaller bats. Ghost bats emerge from their hiding places in caves or abandoned mine shafts soon after sunset, singly or in small groups, and return before dawn. Their hunting technique is to drop on animals from above, enveloping them in their great expanse of leathery wing membranes then killing with bites to the neck and head. The victim is carried to a high perch or back to the cavern where young are being reared. All parts, bones, fur, flesh, teeth and feathers are consumed. Remains in old mine shafts of north-western Australia show that two species of birds most often killed are the Budgerigar and the Owlet Nightjar; its victims are at times as large as the Red-plumed Pigeon.

The Ghost Bat has huge ears that join together above the forehead, a ridged leaf-like nose membrane, and no tail. Fur colour varies, the desert form being ghostly ashy-grey above and white beneath; in the tropical north and north-eastern parts of Australia it is sooty grey. This bat does not occur in southern parts of Australia. The Ghost Bat was formerly believed to be a very rare species, but this was due largely to its extremely secretive habits. During the day it hides in deep fissure of caves and old mineshafts, and is extremely wary. A single young is born each year, and left in the roosting place while its mother hunts at night.

FLYING FOXES AND BLOSSOM BATS

FAMILY *PTEROPODIDAE*

Most of the big bats of this family are fruit and blossom eaters, but some genera contain small partially insectivorous species. Most differ from bats of other families in having comparatively plain or simply furred faces, without the complex folds of skin that form nose-leaf and similar structures (which serve to control emitted sounds) or the complex external ear shapes.

The four largest species, known as 'flying foxes', have wingspans of more than one metre. Apart from their size, they can be recognized by their long-snouted, fox-like faces. They live in large 'camps' containing thousands of individuals, usually situated in rainforest, mangroves or swamp trees, flying out for considerable distances in search of fruit and blossoms. Some of these camps may contain almost a quarter of a million individuals, at a density of between ten and twenty thousand bats per acre.

In addition to the very large flying foxes, this family contains also five species of blossom bats and tube-nosed fruit bats.

Grey-headed flying-fox
Pteropus poliocephulus

This species is very large, grey-furred with a pale reddish-yellow mantle around the shoulders and back of the head, which is a lighter grey

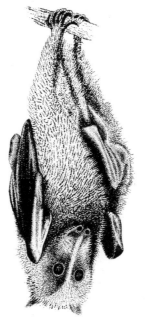

than the body. It inhabits tropical and subtropical coastal areas, occasionally wandering further south, from southern Queensland through New South Wales to Victoria, and occasionally reaching Tasmania.

Although the Grey-headed Flying-fox may occasionally raid cultivated fruit, nectar filled blossoms and wild fruits of the rainforests (such as native figs) make up the major part of their

diet. During the day the flying foxes hang upside down in their camps, which are trees deep in the forest or mangrove swamps. They are capable climbers, using the clawed thumbs of their winged hands, as well as their hind feet.

Usually only a single young is born, and carried each night to the feeding grounds slung beneath its mother's body. When it is older and becomes too heavy it is left at the camp and food is brought back to it.

GLIDERS, LEADBEATER'S POSSUM, RINGTAIL POSSUMS

FAMILY *PETAURIDAE*

There is within this family a considerable variety of tree-dwelling marsupials. Some, the ringtail possums, are slow-moving leaf-eaters; others are extremely active hunters of insects. The family takes its name from the genus *Petaurus*, three medium-sized possums which can glide downwards from tree to tree by means of flight membranes, these being the Sugar Glider, Squirrel Glider and Fluffy or Yellow-bellied Glider. Other groups within this family are the ringtail possums (genus *Pseudocheirus*), the Lemur-like Ringtail *(Hemibelideus)*, the Rock-haunting Ringtail *(Petropseudes)*, the large gliders *(Schoinobates)*, the Striped Possum *(Dactylopsila)* and Leadbeater's Possum *(Gymnobelideus)*.

This family is distributed throughout most of Australia's coastal forests and woodlands, with the majority of species occurring along the east coast where rainforests, eucalypt forests and woodlands provide the greatest variety of habitats. None occurs in the dry interior which is hardly surprising considering that all are dependent upon trees; only one, a ringtail possum, occurs in all of Western Australia south of the Kimberleys. There are about eleven species and numerous sub-species in the family, which is also represented in New Guinea.

Leadbeater's Possum
Gymnobelideus leadbeateri

This beautiful little marsupial is about thirty centimetres in length from its nose to the tip of its long bushy tail; it has a stripe on the face, and looks rather like the common sugar glider but without any gliding membranes. This species was formerly considered to be extremely rare, possibly extinct. Very few specimens of Leadbeaters Possum were ever collected, and these had been taken around 1867, when it was first described and named. In spite of intensive searches in the localities of the original specimens, no trace of the little marsupial could be found, so after more than ninety years it seemed certainly extinct. One factor alone offered hope — the known habitats of this species were in extremely mountainous, heavily forested country, not greatly disturbed by human activities. Here a small marsupial could remain un-noticed, even for as long as ninety years.

Leadbeater's Possum was rediscovered by a party of naturalists using spotlights at night during a survey of the mammals of the Cumberland Valley, only 113 km. from Melbourne. It was glimpsed for a moment climbing through the treetops, showing itself just long enough to be distinguished from the common Sugar Glider. Once the possum's habitat preferences were established, naturalists found that it occurred in many similar places; within parts of its restricted range of distribution in the dense eucalypt forests of south-eastern Victoria it is quite common.

Lemur-like Ringtail
Hemibelideus lemuroides

This ringtail possum is in appearance the most unusual of Australia's ringtail possums, and looks very much like a Greater Glider, although it lacks any gliding membranes. The tail, like that of the similar-sized glider, is fluffy-furred to the tip rather than narrowly tapering like the tails of other ringtail possums; the face is remarkably similar to that of the Greater Glider, and it has much the same dusky brown soft woolly fur. Therefore a separate genus was created for it (most other Australian ringtails being in the genus *Pseudochurus*); *Hemibelideus* means 'semi-glider'.

In keeping with the Glider-like appearance, the Lemur-like Ringtail is much more active in the treetops than the other ringtails, which move slowly and deliberately, always maintaining their grasp with at least the tail and one foot. The Lemur-like Ringtail makes bold downward leaps among the branches; it has a very slight beginning of a gliding membrane, a strip of furry skin about two and a half centimetres wide along each flank.

This possum is named for its facial likeness to the lemurs, which are monkey-like primates of Madagascar, Africa, India and Malaya. It is confined to the dense rainforests of the Herbert River's mountainous headwaters, and the Atherton Tableland region of north-eastern Queensland.

Sugar Glider *Petaurus breviceps*

A medium-sized, superficially squirrel-like possum with a very long fluffy tail; head and body together eighteen centimetres, tail twenty centimetres. The fur is a delicate ash-grey, creamy white below with a black dorsal stripe extending back from between the big bulging black night-seeing eyes. The Sugar Glider, and the almost identical but slightly larger Squirrel Glider *(P. norfolcensis)*, glide from tree to tree by means of their flight membranes. These are lateral extensions of the body skin which, when all four limbs are outstretched, extend tightly between front and hind limbs. When the glider leaps from the top of a tree it has the appearance of a rectangular parachute, at first falling almost vertically. As its velocity increases, giving more lift to the 'wings', the steep dive levels out, becoming almost horizontal as the little marsupial nears the ground, and terminating in an upward swoop as it comes into a vertical position to land and cling on the bark, low down on the trunk of another tree.

Sugar Gliders are very active, fast-moving in their nocturnal hunting for insects, nectar, fruits, and the sweet gum that exudes from the bark of some eucalypts. They inhabit eucalypt forests and woodlands of northern, eastern, and south-eastern coastal Australia, living in small communities or family parties, probably made up of the young of several seasons. The two young at first remain in the pouch for several months until they become too heavy to be carried by the mother on her nocturnal flights, when they are left in their tree-hollow nest.

Herbert River Ringtail
Pseudocheirus herbertensis

The ringtails are slow-moving possums distinguished by their very long tapering tails which are carried curled in a ring when not being used to grip branches as an aid to climbing. Their ears are very short, the eyes large, rounded and bulging out from the face in a manner common among the nocturnal marsupials.

The Herbert River Ringtail or Mongan is restricted to the mountain rainforests of north-eastern Queensland. It was first discovered in the wild country around the headwaters of the Herbert River, near the Atherton Tableland.

This species is probably the most attractive of the ringtail possums, having rich reddish or

chocolate-brown fur on the upper parts, changing abruptly to pure white on the under surfaces and parts of the limbs. The big eyes are a bright brick red.

Like other ringtails this inhabitant of the dense rainforests is herbivorous, living on leaves and native fruits.

HONEY POSSUM

FAMILY *TARSIPEDIDAE*

This family contains just one species, the unique Honey Possum. Its body, snout and limbs have become so extremely modified that little is known of its ancestry except that it has no close relatives among marsupials that inhabit Australia today. Its nearest but distant relatives are the similar-sized pigmy possums.

The most extreme of its adaptations for a diet of pollen, nectar and the tiny insects found in flowers are to be seen in the modifications of the head, the snout becoming in function an equivalent of the long bill of honeyeating birds, while its tongue also has become similar to the brush-tipped tongues of these birds. The Honey Possum's snout has become lengthened to almost a trunk or proboscis. There are flanges along its lips which overlap to form a tubular channel through which nectar, pollen and microscopic insects can be sucked up. Its slender tongue is covered with fine bristles with a tuft at the tip, and it can dart out twenty-five millimetres beyond the tiny animal's nose. It is all but toothless, these having degenerated after millions of years of such soft foods.

As the single species is found only in south western Australia, the family likewise must be endemic to that region.

Honey Possum *Tarsipes spenserae*

Known also by its aboriginal name, Noolbenger, this mouse-sized marsupial has a head-and-body length of about seven and a half centimetres, and a tail of eight or nine centimetres. It has three well defined stripes down its back, a very long slender whip-like tapering prehensile tail with a naked area near the tip, and an extremely elongated almost tubular snout. It is confined to the south-western corner of Western Australia from the Murchison River on the west coast to Esperance on the south coast, usually inhabiting sandplains and shrub-heathlands.

Many of the commonest wildflowers of this habitat are deeply tubular (grevilleas, lambertias) or like stiff brushes (banksias, bottlebrushes). Their nectar can be reached only by an animal equipped, like the long-billed birds or the Honey Possum, to reach into such narrow spaces. In this region, on the heathlands, there are at any time of the year some wildflowers in bloom, autumn and winter included. These little marsupials wander (possibly in small colonies) according to the flowering, at one season being found on the bottlebrush shrubs, another time among the banksias. During the day Honey Possums hide in nests of grass and fur, usually built in dense foliage, such as in the tops of grass-trees; they are occasionally found curled up in old nests of small birds. The young are probably kept in these nests after leaving the pouch.

HORSESHOE BATS

FAMILY *HIPPOSIDERIDAE*

These small bats have prominent horseshoe shaped skin structures around the top of the snout. The nose-leaf structures probably play an important part in the bats' radar-like echo-location of objects in their flight paths. Possibly the nose-leaf channels the emitted ultra-sonic sound in a beam ahead of the flying bat. In Australia four groups of bats have nose-leaves but none are more prominently developed than in the two families of horseshoe bats (the other family being the Rhinolophidae).

The complex nose-leaf structures are made up of three main parts. There is a lower part which is large and horseshoe shaped; there is a central part around the nostrils which may carry a sound-projecting organ, and there is an upper or hindmost part, towards the forehead. The details of these are useful in identification of the species.

The fur of horseshoe bats is often colourful. They are usually greyish-brown but some species have a bright rufous or even orange colour phase; they may change from grey-brown to orange with age.

Diadem Bat *Hipposideros diadema*

This species is also known as the Large Horseshoe Bat, being distinguished from others by its greater size. In its nose-leaf structure the upper part is distinctly wider than the lower, and divided into four depressions by ridges giving a diadem pattern.

A cave-dwelling bat, the Diadem is found in north-eastern Queensland from Cardwell northwards to Cape York. Elsewhere it has a wide range which includes New Guinea and extends to south-eastern Asia. Australia's Diadem Bat is a distinct subspecies. Little is known of its habits except that it hunts insects after dark. It is thought that the elaborate nose structures are part of its sensory radar-like equipment which probably locates the prey in total darkness.

KANGAROOS AND WALLABIES

FAMILY *MACROPODIDAE*

The marsupials of the macropod family can readily be distinguished from all others by a number of obvious characteristics. The hind limbs are very large in proportion to the rest of the body, and adapted for a two-footed leaping action. The feet are very long (hence the term 'macropod') with the fourth toe very large and heavily clawed. The other toes are smaller, the second and third toes very small and joined together except for the claws. These twin-clawed, fused-together toes on each hind foot are used like a comb in scratching through the fur. The typical macropod forelimbs are much shorter and far more lightly built. The tail is large, heavy, and used as a balance in the leaping fast travel, or as a kind of auxilliary limb when standing or moving slowly in quadrupedal manner.

The macropods have teeth modified for grazing, with two very large scissor-like incisors at the front which press up against a leathery pad behind the upper front incisors. Except for one species (the Musky Rat-kangaroo) the macropod family has evolved a digestive system parallel to that of the ruminants of other continents, the stomach being large and sacculated, so that bacteria and protozoa contained in large numbers can carry out a preliminary digestion of the bulky foliage or herbage food.

The larger macropods are commonly known as kangaroos, wallaroos and euros, the smaller as wallabies, rat-kangaroos and potoroos. There is no distinct differentiation between these common names — they result mainly from traditional popular usage. This is a large family of about forty-six Australian species; it is represented also in New Guinea.

Agile Wallaby *Macropus agilis*

The Agile Wallaby, also known as the Sandy Wallaby and River Wallaby, is the most abundant large wallaby of northern Australia. Its distribution covers the entire tropical north, from eastern coastal Queensland through the 'top end'

of the Northern Territory to the Kimberley district of Western Australia. It is most often seen in river flood-plains country and the blacksoil plains, where the tropical grasses grow fast and tall beneath scattered eucalypts after the summer wet season. The Agile Wallaby feeds at night, and by day shelters in the jungle-like patches of monsoon forest or other dense scrub.

This Wallaby has bright sandy or golden-brown fur with very distinct white hip and cheek stripes, and short ears. It has a head-and-body length of about one metre, and a tail a few centimetres shorter.

As with many of the macropods, the difference in size between the sexes is conspicuous. With age the males increase steadily in height and bulk, while the females remain far smaller. A number of races have been described, based purely upon slight differences of fur colour, from dull sandy-brown around Darwin, and a brighter yellowish-brown on Cape York, to reddish-sandy in the West Kimberley.

Bennett's Wallaby *Macropus rufogriseus*

Next down in size after the big kangaroos, wallaroos and euros are the large wallabies, most of which belong to the same genus, *Macropus*, as the majority of the kangaroos. Whether they are called wallabies or kangaroos means nothing except a rough indication of size.

Macropus rufogriseus is a very widespread large wallaby, and consequently has several distinct geographical races in various parts of Australia. It was originally named from a specimen taken on King Island in Bass Strait, and now known as Bennett's Wallaby. It is of slightly smaller, stockier build than the two other sub-species, and the reddish fur of its shoulder and rump can only just be distinguished from the grey-brown of the remainder of the body. One sub-species almost identical, inhabits Flinders Island in Bass Strait, and occurs in Tasmania, where it is known as the Brush Kangaroo, or as Bennett's Wallaby.

The third form is that occurring on the south-eastern coastal parts of the Australian mainland. It has much more conspicuously reddish shoulders and rump, and is generally known as the Red-necked Wallaby.

This large and gracefully built Wallaby prefers the forest undergrowth, browsing on foliage rather than venturing out onto open plains to feed on grass. Although driven from much of its original territory, it still survives in the more rugged parts.

Brush-tail Bettong *Bettongia penicillata*

Formerly very widespread in western New South Wales, western Victoria, through South Australia and Central Australia to Western Australia, the Brush-tailed Bettong or Woilie is almost certainly extinct everywhere except in the south-west corner of Western Australia. It is a small marsupial, not much more than rabbit-sized, with a head-and-body length of 35 centimetres and a tail of 30 centimetres. Its tail is unusual among the macropods in that it is prehensile, and can be curled around bundles of sticks or grass to carry these materials to its nest

site. The tail is crested with long black hairs towards the tip, forming a brush.

This nocturnal, herbivorous, rat-kangaroo hides during the day in a nest of grass, sticks and other forest debris, built in a small hollow scratched out beside a low bush. These nests are extremely well concealed, being, externally, just like any other clump of the rather dry looking herbage. Grass is drawn across the entrance by the occupant within, when settling down for the day.

Related species of these miniature kangaroos have suffered equally heavy losses since settlement of Australia. The Eastern Bettong *(Bettongia gaimardi)* which has a white-crested tail, has not been recorded on the mainland since 1910 but fortunately remains common in Tasmania. The Boodie *(B. lesueur)*, which was once common from New South Wales to Western Australia, is now but a rare inhabitant of remote central Australia, and common only on a few offshore islands.

Brush-tailed Rock-Wallaby
Petrogale penicillata

Rock-wallabies have become well adapted for a life among the boulders, caves and cliffs of mountain ranges and rocky hills in many parts of Australia, both coastal and inland. Their rock-dwelling habits have led to the development of certain features which are quite distinctive. The hind feet are very well padded, with the soles of the feet roughly granular to prevent slipping on polished rock surfaces. Their tails are long and slender, usually brush-tipped or tufted to serve as a rudder while the wallaby is airborne on long leaps across crevices and between boulders, and to act as a balancer on narrow ledges and ridges.

The Brush-tailed Rock-wallaby has many variations of fur colour and markings because of its wide distribution, which is broken up into pockets where suitably rocky country occurs; isolated populations have tended to develop slight differences. It has a head and body length of about seventy-five centimetres, and a tail of sixty centimetres. The long tail is dark, and

bushy towards the tip; it does not appear to taper like the tails of most other wallabies and kangaroos. This species occurs throughout Australia in hilly and rocky localities, but is absent from Tasmania; there are about twelve sub-species within this population.

It is probable that this wallaby's exceptionally wide distribution, from cold south-coastal islands to tropical desert regions has been made possible by the subdued and constant temperatures that prevail in the rock crevices and caves where it shelters during the day. Here it may be as much as 12°C cooler than the outside shade temperature in summer.

Desert Rat-Kangaroo
Caloprymnus campestris

This extremely rare inhabitant of the most arid desert country has had a remarkable history of discovery, loss and rediscovery. Always extremely rare, it has not been seen for many years and may possibly have joined the list of permanently lost Australian native animals. It inhabited the desert flats between sand dunes, and the stony plains bordering the Simpson Desert, near where south-western Queensland and the north-western corner of South Australia meet.

This desert animal, besides being one of the smallest of the kangaroo family, is one of the most beautiful in colouring. The fur as a whole is sandy-buff, which is an admirable protective colouration for the desert habitat. The middle of the back is of grizzled texture owing to the fur having within its depths five distinct colour bands including shades of cinnamon, pale yellow, and very dark brown. This gives the fur a remarkably variegated effect when ruffled or separated. The sides of the body are a rich sandy-yellow. This rat-kangaroo has a rounded head and blunt face, and the front limbs are very small, only one-third the length of the long hind legs.

Unlike many desert animals, it has not adopted the burrowing habit. Its daytime nest is a simple hollow about ten centimetres deep, covered with a lacework of twigs and grass stems, usually placed under the shade of a small shrub. The Desert Rat-kangaroo was first named by John Gould in 1843, lost for the next ninety years, and rediscovered in 1931. Since then, contact with that small desert population has again been lost.

Great Grey Kangaroo   *Macropus giganteus*

The Great Grey or Forester rivals the Red Kangaroo in size, with a head-and-body length of 1.5 metres and a tail of about one metre. The fur is short, silvery grey on both male and female, and without any noticeable lighter or darker markings. The snout is hairy between the nostrils, unlike the naked-nosed wallaroos. Females are conspicuously smaller than the fully-grown males. The Great Grey inhabits eastern Australia from north-eastern Queensland through New South Wales to Victoria and extends inland to the western districts of New South Wales. It overlaps the range of the Red Kangaroo, but keeps to the dense scrubs and forests leaving the

open woodlands and grasslands to the Red. The Tasmanian Grey is a subspecies.

Like the Red Kangaroo, the Grey is a gregarious animal, usually seen in groups. It inhabits forested country, sheltering during the day in dense vegetation or among boulders, and emerging at evening and early morning to feed in grassy clearings.

There are in Australia various other grey kangaroos, which are all closely related to each other, and very similar in appearance; they are more distantly related to the Great Grey. These are the Kangaroo Island Kangaroo *(Macropus fuliginosus)*, the sub-species known as the Black-faced Kangaroo or Mallee Kangaroo, and the Western Grey Kangaroo of south-western Australia.

Karrabul   *Onychogalea unguifera*

The three species of distinctively marked nail-tailed wallabies have a most unusual characteristic — hidden in the dark hair at the tip of their long slender tails is a small dark horny nail, rather like a finger nail. Their noses are strongly convex in profile, giving a 'Roman-nosed' appearance which is characteristic of all species. Nail-tails travel in a curious head-down attitude, the fore-parts of the body being held as close to the ground as the long tail, while leaping leisurely along. The tails are very long and slender and the incisor and premolar teeth are different from those of other wallabies. All species have distinctive white cheek, shoulder and hip stripes, the exact positioning of which is the principal means of identifying each species, which otherwise are very similar.

The Karrabul or Northern Nail-tail Wallaby has the shoulder stripe beginning level with the armpit and extending down onto the chest. It is a common species across tropical northern Australia from inland north-eastern Queensland to the Kimberleys, in savannah woodlands. The Karrabul is a little larger than the other species, and has a nose-to-tail length of around 1.25 metres.

The Crescent Nail-tailed Wallaby *(O. lunata)* once inhabited south-western and southern central Australia, but long ago vanished from inhabited regions; it is now a very rare inhabi-

---

Lemur-like Ringtail   *Hemibelideus lemuroides*
Named for the likeness of its face to the lemurs of Madagascar, which are monkey-like primates, the Lemur-like Ringtail has as much in common with the Greater Glider as it does with other ringtails, and is thought to be a link between those groups. It is also known as the Brush-tipped Ring-tail, a reference to the bushy glider-like tail.

Honey Possum   *Tarsipes spenserae*
The tiny Honey Possum has developed, over a period of many thousands of years, a unique relationship with the many flowering plants of south-western Australia. It is completely dependant upon these large wildflowers for its nectar food, and its mouth and snout have gradually evolved into a sucking tube. The flowers benefit from the cross-pollination accomplished by the little marsupial.

Bennett's Wallaby *Macropus rufogriseus*
An inhabitant of King Island, in Bass Strait, Bennett's Wallaby is a large yet gracefully built species, an inhabitant of forests, and a browser of scrub foliage rather than grazer of grass. A very similar sub-species, the Rednecked Wallaby, occurs on parts of the Australian mainland.

Agile Wallaby *Macropus agilis*
Wallabies are, as a group, very little different from the great kangaroos; their smaller size is the feature that sets them apart from their big relatives, and that too is no more than a consequence of their environmental specialization. Whereas the kangaroos are inhabitants of open spaces, the wallabies prefer forest

undergrowth, or the scrub along rivers in open country. The Agile Wallaby of tropical northern Australia is also known as River Wallaby and Jungle Kangaroo because it hides by day in dense patches of tropical jungle or other dense vegetation (which in many places occurs along rivers and around swamps) and ventures out at night to feed in the lush tall grass.

tant of the most remote desert regions. The third species, the Bridled Nail-tail Wallaby, once inhabited the interior of New South Wales and southern Queensland; there has been no reliable record of it for thirty years.

Long-nosed Potoroo *Potorous tridactylus*

This small rat-kangaroo, which has a head-and-body length of forty centimetres, and a tail of twenty two centimetres, superficially resembles a bandicoot. The ears are very short and rounded, tail short, tapering and prehensile. It hops along with its body held horizontal and close to the ground. The tail is often white-tipped, and the body fur may be dark grey or fawn.

Like other potoroos, this species (also known as Gilbert's Potoroo and Long-nosed Rat-kangaroo) is nocturnal, and feeds on roots and tubers which it digs up with the long claws of its forefeet. It keeps to areas where the ground cover is dense, such as tall tussocky grass and ferns beneath eucalypt forest or woodland. During the day potoroos sleep in grass nests constructed in dense herbage or in small hollows under bushes or grass. The slightest noise disturbs them and they dart away beneath the undergrowth.

This species still survives in coastal south-eastern Queensland, coastal New South Wales. It also occurred in north-eastern Victoria and the south-western corner of Western Australia. The Western Australian subspecies, once listed as a separate species *(P. gilbertii)*, is probably now extinct; elsewhere the species has vanished from most of its previous haunts, probably owing to destruction of vegetation and introduction of predators. In north-eastern New South Wales a survey showed that where it lived in tall woodlands next to rainforest, it was not found in the rainforest.

Lumholtz Tree-kangaroo
Dendrolagus lumholtzi

Australia's two species of tree-kangaroos have come from New Guinea in the distant past at the time when that island was linked with Cape York. They have since then become sufficiently different to be considered species distinct from their New Guinea relatives. Both species have remained confined to that part of Australia which most closely resembles the original environment.

Tree-kangaroos were once ground-dwellers but returned again to the tree-dwelling way of life of the Kangaroo family's ancestors. They still have the basic kangaroo shape, but it is now modified for tree-climbing. The hind feet have become shorter and wider with roughened

padded soles for a non-slip grip. The sharp claws are stronger and more curved. The fore-limbs are much heavier and more powerful than those of ground-dwelling kangaroos, and have large strong hands and claws. The very long tails of the tree-kangaroos have never developed any prehensile grip, but are useful in a different way, having a thick fur brush towards the tip serving as something of a rudder on long leaps from limb to limb, and as a balancer as they walk along the branches. Tree-kangaroos show incredible agility, and there are records of them jumping ten or fifteen metres to the ground and landing, cat-like, on all fours, quite unharmed.

The Lumholtz Tree-kangaroo or Boongary is a large marsupial, with a head-and-body length of sixty-five centimetres, and a heavy tail that is equally long. Its face, hand and feet are dark, contrasting with the grey back and very pale belly fur. It inhabits the mountainous rainforests of the Cairns and Atherton Tableland region. The second species, the Dusky Tree-kangaroo *(D. bennettianus)*, is found further north, in mountain rainforests between Cooktown and Daintree, on Cape York.

Musky Rat-kangaroo
Hypsiprymnodon moschatus

The single species of this genus is considered to be the most primitive member of the entire kangaroo family. It still shows traces of this family's possum-like tree-dwelling ancestors in its hand-like hind feet with their opposable thumb-like big toes, a scaly rather than furred tail, and teeth suited to a mixed diet instead of the purely herbivorous dentition of all others of the family.

The Musky Rat-kangaroo is a small bandicoot-like animal about forty-five centimetres in length including the rather short scaly tail. Unlike other kangaroos and rat-kangaroos, its front and hind limbs are of almost equal length, and it runs on all four instead of hopping along on the hind legs. It is also unique among kangaroos in that it is partly insectivorous, obtaining these and other small forms of animal life by scratching like a bandicoot among the debris on the floor of the tropical rainforests which it inhabits. It is confined to coastal north-eastern Queensland, between Townsville and Mossman.

Parma Wallaby *Macropus parma*

One of the rarest of Australian wallabies, the Parma or White-fronted Wallaby was long thought to be extinct, as none had been seen since 1932. It seems that the Parma was never a common species, even at the time of first settlement, when it lived in the rainforests around Lake Illawarra, New South Wales. It was rediscovered in 1965, when close examination of some wallaby skins sent from New Zealand showed that wallabys remarkably similar to the Parma lived there. Further investigations revealed that the small wallabies which occurred in great numbers on the small island of Kawau, off the coast of New Zealand, were indeed Parmas. They had been taken there long ago by Sir George Grey, a governor of New Zealand and once an Australian explorer with a deep interest in Australian animal life.

Not only did Australia's extinct Parma still exist on this New Zealand island, but in such numbers that the wallabies were being shot by the thousands because they were damaging young trees in pine plantations. Small numbers have now been returned to their homeland, as the beginning of a re-establishment programme. By remarkable coincidence the Parma was later found to survive still in dense forest country near Gosford.

The general colour of this little wallaby is dark brown, with a faint dorsal stripe down to the mid back. The upper lip, throat, chest and belly are white. Its former habitat was the rainforest and scrub of eastern New South Wales.

Quokka *Setonix brachyurus*

This well-known inhabitant of the south-west of Western Australia is one of the smallest of wallabies, being only about the size of an ordinary cat. It is a rather dumpy, stumpy-limbed little wallaby, with a very short face, and rounded ears upon the top of its head. Until the mid 1930s the Quokka was very common in the south-west, where it occurred in swampy thickets. Quokka shooting was a popular sport. Today it is very rare on the mainland, the last surviving colonies being in densely vegetated swampy-valleys of the Darling Ranges near Perth.

Fortunately the Quokka's distribution included two offshore islands, Bald Island, off the mid south coast east of Albany, and Rottnest Island, westwards of Perth. The latter is a very popular summer holiday resort, and here the little Quokkas have become so tame that they are to be seen in daylight in and around the settlement, and soon become familiar visitors to most campsites.

The Quokka, so numerous and close to Perth at Rottnest Island, became a convenient laboratory animal for zoological studies; most of the foundation work on the biology of marsupials was done with Quokkas. One of the notable discoveries was that of the delayed development of any embryo in the womb, that occurred if there was already a joey being carried in the pouch. (This feature of marsupial reproduction is described for the Red Kangaroo).

Red-bellied Pademelon
Thylogale billardierii

The small wallabies of the pademelon group (genus *Thylogale*) are related to the rock-wallabies, and dwell in thick scrub, such as rainforest. Generally, the term 'pademelon' applies to macropods smaller than kangaroos and the majority of wallabies, yet still considerably larger than hare-wallabies or rat-kangaroos. Apart from their small size, pademelons are distinguished by their comparatively short, sparsely furred, thick, un-tapered tails, a slightly more stocky, shorter-legged build, and distinctive dental characteristics.

The pademelons live among thick scrub, dense undergrowth of forests, tangled long grass and the dense low scrub of swampy country. Here they push through the tangled vegetation to make tunnel-like runways, which often indicate the presence of these animals when they themselves cannot be found. At dusk and sunrise they emerge from concealment to graze on open grassy areas, usually remaining very wary and not venturing too far from cover. They feed upon grasses, foliage and shoots.

The Red-bellied Pademelon is abundant in Tasmania and evidently occurred in south-eastern Australia in the earliest days of European settlement. It is a very gregarious species, occurring in large communities which are like warrens with their interconnecting runways through the undergrowth. Because of its large numbers it is considered a pest in some farming areas, and great numbers are killed for their fur.

Red Kangaroo   *Megaleia rufa*

Also known as the 'Plains Kangaroo', this very large heavy marsupial, the largest of all kangaroos, has a head-and-body length of 1.6 metres or more, and a tail of about one metre in length. Its fur is short, dense and woolly; although most commonly red, the colour varies in some regions. Usually, the big males are entirely dull red except for distinctive and unvarying black and white markings on each side of the muzzle, and the pale tail tip, while the considerably smaller females are blue-grey. Sometimes however the males can be entirely grey, or grey-bodied with the head red, while females are generally of a different colour to the males of the same locality. However the muzzle markings do not vary.

This kangaroo is very widespread across almost the whole continent except for the far north, the forested and mountainous eastern coast, the south-east and the south-west. Its abundance varies with degree of settlement and persecution, and seasonal conditions. The habitat is principally of semi-arid plains, where droughts are of frequent occurrence. In addition to being able to withstand great stresses from heat and shortage of water, these kangaroos cease to breed when conditions become extremely adverse. However when environmental conditions become favourable the building up of the kangaroo population is aided by the ability of the females to carry a reserve embryo in the uterus at the same time that there is a joey in the pouch. The development of that second young is stopped at an early stage as long as the older one remains in the pouch, but it is born within a day of the first permanently vacating the pouch. So while the kangaroo can produce only one young at a time, these are as closely spaced as possible.

Rufous Rat-kangaroo
Aepyprymnus rufescens

This is the largest of the little rat-kangaroos, with a head-and-body length of slightly more than fifty centimetres, and a tail of about thirty-seven centimetres. The head and body are rufous-grey, the ears fairly long, tail tapering and neither crested nor prehensile. It has a whitish hip-stripe, black on the back of the ears, and the muzzle tip is hairy. There is a mingling of stiff silvery grey hairs throughout the rich brown fur which gives the grizzled overall rufous-grey colour.

The Rufous Rat-kangaroo, the only species within this genus, inhabits eucalypt woodlands and forests which have a ground cover of dense tall grass. It is not a gregarious animal and is usually seen singly or in pairs. This species, known in some areas as 'kangaroo-rat', builds nests in tall tussocks of blady-grass or similar dense low vegetation. These nests are occupied during daylight hours in all seasons by individuals of both sexes; except where females have a large young with them, there is only one animal to each nest. If disturbed they dash out of the nest and do not appear ever to come back to that particular nest. Built under a dense tussock, often against a log or under fallen branches, the nest consists of a basin-shaped hollow scratched into the ground, into which a thick-walled spherical nest mainly of dry grasses, with a side opening, is built. Nesting material is carried in the curled tip of the tail.

The Rufous Rat-kangaroo is still quite common in some hilly parts of its range of distribution in eastern Queensland, north-eastern New South Wales and north-eastern Victoria.

Spectacled Hare-wallaby
Lagorchestes conspicillatus

The name 'hare-wallaby' was given when, in the early days of the exploration and settlement of Australia, these rabbit-sized wallabies were abundant.

If their nesting places under tussocks of grass or low bushes were approached they would dash out with hare-like abruptness and speed. However the species which inhabited eastern Australia is probably extinct, the last having been seen in 1890; this was the Brown Hare-wallaby (*Lagorchestes leporoides*).

Three species still survive because they occur in remote areas on islands which give them protection from the introduced cat and fox. The Banded Hare-Wallaby (*Lagostrophus fasciatus*) is common on Bernier and Dorre Islands in Shark Bay, Western Australia, but no specimen has been taken on the mainland since 1906. The Western Hare-wallaby (*Lagorchestes hirsutus*) occurs on these same islands, and inhabits spinifex grasslands in the far northern interior of Western Australia where it may still survive in the remote country traversed by the old Canning Stock Route.

Like these other island refugees, the Spectacled Hare-wallaby has its stronghold on an island — Barrow Island, off the north-west coast of Western Australia. It also occurs in the remote northern parts of the interior of Western

Australia and the Northern Territory. It is probably the only member of the once abundant hare-wallabies still reasonably secure on mainland Australia. Although all species have orange rings of fur around the eyes, the feature is far more noticeable on this species, hence the name 'spectacled'.

Whiptail Wallaby   *Macropus parryi*

The name 'whiptail' refers to the great length of the slender tail, which is as long as head and body together. It is also known as the Pretty-face Wallaby, Grey Flier or Blue Flier. The long rather shaggy fur is grey-brown, darker brown about the eyes and on the snout, with white cheek stripes, and white hip stripes. The tail is paler, with a dark tip. The Whiptail inhabits open grassy eucalypt woodlands, usually favouring the hill slopes and summits of the coastal ranges. It occurs in eastern Queensland and north-eastern New South Wales.

Whiptails are probably faster than most other wallabies, and more inclined to be out in the daylight, small mobs feeding on the open grasslands as late as ten in the morning and reappearing after four in the afternoon. The heat of the day is spent under the shade of clumps of trees or scrub. No clearly defined breeding season has been noted.

This wallaby is more gregarious than most others, usually being in groups of four up to twelve individuals, occasionally less or more. In a survey of the mammals of the upper Richmond and Clarence River area of north-eastern New South Wales, it was found that this graceful wallaby occurred in suitable habitats in most parts of the region, where it was common or of regular occurrence. Where hill slopes had been partly cleared for cattle grazing, thus creating habitat suiting this species, its numbers had increased, and it was often to be seen in late afternoon feeding out on open areas, and grazing on the more nutritious introduced grasses.

KOALA

FAMILY   *PHASCOLARCTIDAE*

The single famed and unique species within this family is one of the most specialized of the herbivorous marsupials, never coming to the ground except to cross from one tree to another, and accepting as food the leaves of only a few species of eucalypts. It has become highly adapted for this way of life. Although lacking the prehensile tail which makes possums such skilful climbers, the Koala compensates with vice-like grip of heavily-clawed hands and feet.

Of all Australian marsupials, the Koala appears to be most closely related to the wombats. Once it was thought that the Koala was a sort of over-specialized member of the possum family, however it cannot be linked with possums except by superficial characteristics. Although similarities of foot structure and dentition might be regarded as indicating relationship with ringtail possums, these are merely the influence of a tree-dwelling, leaf-eating way of life rather than any ancestral connections. Characteristics suggesting a common ancestry shared by Koala and wombats include the rearwards opening pouch, cheek pouches (rudimentary in wombats) and lack of significant tails. While a rearward opening pouch is perfect for the burrowing wombats, for the tree-climbing Koala it then becomes downwards-opening, and the young is prevented from falling out only by a tight muscular entrance; this feature suggests ground-dwelling ancestors.

Koala *Phascolarctos cinereus*

The Koala is quite a large animal, growing to length of eighty centimetres, and of very bulky shape. It is tail-less, has a naked bulbous snout, and has long arms with the first and second fingers opposable to the other three. It is not only specialized in its arboreal life, but in its selection of certain types of eucalyptus leaves, and it seems that the Koala needs a varied diet selected from among these. This may be because some, such as the Manna Gum, produce poisonous prussic acid in new leaves at certain times. In Victoria the principal food trees are the Manna Gum, Messmate, Mahogany, Peppermint and Swamp Gum, while in New South Wales the Brush Box, Sydney Blue Gum and Forest Red Gum are eaten. A remarkable feature of the Koala's anatomy is its so-called 'appendix' which

attains a length of more than two metres. It is in no way comparable to the degenerate and probably functionless human appendix, but is really a great caecum or extra prolongation of the intestines to aid the digestion of the bulky leaf diet. It seems that Koalas rarely drink water, this being provided by the foliage eaten. The name Koala may derive from a similar sounding native word meaning 'no drink'.

This animal is a summer breeder; the young are born about one month after mating, and are carried in the pouch for a further five or six months. Towards the end of this time the young enters and leaves the pouch at will and often rides on the mother's back. The original distribution included Southern Queensland, eastern New South Wales, Victoria and South Australia, the preferred habitat being eucalypt forests and woodlands.

MARSUPIAL-MOLES

FAMILY *NOTORYCTIDAE*

The single species of this remarkable family represents a remarkable example of convergent evolution. Although it is not even distantly related to moles of other continents (except that all are mammals) the Australian animal has followed the same evolutionary path, marsupial and placental moles now being outwardly very much alike. In general form the marsupial-mole is very like the African Golden Moles (family *Chrysochloridae)*. The close similarity results from the development of the body shape best suited to movement through the soil with least resistance, limbs best able to provide subterranean propulsion, the loss or reduction of senses of sight and hearing and probably their replacement by heightened senses of smell and touch. So specialized is the marsupial-mole that its relationship to other marsupial families remains in doubt, its extensive modifications obliterating features which might have indicated its origins. Some features of dentition and structure of the feet suggest a possible common origin with the bandicoot family. That it is a marsupial is proven by the fact that it has a pouch. Two separate populations are known, inhabiting sandy semi-desert country of southern-central and north-western Australia; the north-western form is sometimes listed as a separate species, *N. caurinus.*

The Marsupial-mole *Notoryctes typhlops*

A small animal, about fifteen centimetres in length, with a smoothly streamlined shape, pointed at both ends, the head 'flowing' onto the body with no narrowing for a neck. The conical nose is covered with a hard horny shield, the eyes have degenerated and entirely disappeared being of no use to an animal almost permanently beneath the soil, and the tail is reduced to a small leathery remnant.

The limbs are short and powerful, with five digits on hands and feet. These show great adaptation for burrowing, the claws of the third and fourth fingers being enormously enlarged and scoop-like to serve as pick and shovel. The fur is

fine and silky, with a velvety, almost iridescent sheen, and varies in colour from almost white, to a deep gold or orange. The pouch, as would be expected, is of the backward opening type.

The marsupial-mole tunnels along about five to ten centimetres below the surface in soft sand-dune country, the burrow falling in behind it, and generally leaving a trace on the surface. The few specimens obtained have mostly been found on the surface after rain. Earthworms and subterranean insects make up the diet; very little is known of breeding except that one young has been found in a pouch.

NATIVE MICE AND NATIVE RATS

FAMILY *MURIDAE*

This large family, which is also widespread outside Australia, contains more than one-third of the rodents of the world. All members of this family have two prominent upper incisors which are continuously growing and have hard enamel on their front surfaces only. Australia, at the time of its discovery by European man, contained almost fifty species of native mice and rats, all clean and attractive bush-dwellers. Some of these have been so long isolated on this island continent that they are considerably different to any of the rodents of other continents, while others found their way here not so many thousands of years ago and still closely resemble the mice and rats of other lands.

Within Australia the native mice and rats have adapted to suit the great variety of habitats available. Naturally the earliest arrivals show the greatest degree of adaptation. The hopping-mice have adapted to arid conditions, and have acquired a miniature kangaroo shape. At the other extreme, our native water-rats have become specialized for an aquatic life in Australian streams. A great many of the genera and species within the family Muridae are unique to Australia.

Many of the species of native rats and mice are difficult to identify, most genera and species being based on features of skulls and teeth requiring expert examination. However the locality of any particular specimen can be a useful guide to its possible identity.

Golden-backed Tree-rat
Mesembriomys macrurus

The little-known Golden-backed Tree-rat is one of the most beautiful of Australian native rodents. Although not often seen, it is still quite common in the remote parts of tropical northern Australia, from the West Kimberley to the Northern Territory, where it is found in woodland savannah country.

The native rats of this genus are often known as rabbit-rats because they are much larger than ordinary rats, and have long ears, narrow feet and long bushy tails. The most conspicuous feature of this species is the golden brown stripe down the back from forehead to rump, contrasting strongly with the buffy-grey sides and creamy white underparts. It is about fifty-two centimetres in length, of which the long white-tipped tail accounts for thirty centimetres.

This rat and the one other species of the genus (the Black-footed Tree-rat) are tree-climbers and apparently live in hollow trees, descending to the ground to feed.

Little Native-mouse   *Pseudomys delicatulus*

Some of Australia's rodent native mice (not to be confused with the marsupial mice) are very similar to the ordinary introduced house mouse in appearance. Most are difficult to identify, differences being more in features of skulls and teeth rather than obvious external characteristics. The native mice of the genus *Pseudomys* (which means 'false mouse'), although like ordinary mice, are shy bush-dwellers. Their habits and distributions are for the most part not well known.

The Little Native-mouse is about the same size as a small specimen of ordinary mouse. It is greyish-fawn above, abruptly changing to white below. The tail is about the same length or longer than head and body. It is one of the smallest and most delicately coloured of our native mice, and remains common in some of the more remote areas, particularly the north-west Kimberley and on the Cobourg Peninsula in the Northern Territory. Its recorded distribution was northern Australia from the west Kimberley district of Western Australia to the eastern coast of Queensland south to about Rockhampton.

Shaggy Rabbit-rat   *Mesembriomys gouldii*

This is one of Australia's largest rodents, with a head-and-body length of around thirty centimetres and a tail of thirty-three centimetres. It has a long-haired, shaggy coat of fur, a large white-tipped brush-tail, black ears and feet. The general body colour is yellowish brown, but longer black guard hairs overlying the shorter yellowish coat give a more sombre and shaggy appearance.

The Shaggy Rabbit-rat is found in tree-covered country in northern Australia, from Cape York Peninsula and Melville Island westwards through coastal parts of the Northern Territory to the northern Kimberley region. It makes its nest of leaves and bark within hollows of trees. An extremely savage temper and severe bite have been commented upon by many who have captured specimens of this powerful and agile rat, which is armed with large chisel-like incisor teeth. It is still common in the remote northern regions which it inhabits.

Southern Bush-rat   *Rattus fuscipes*

The Southern Bush-rat lives only in natural bushland, avoiding human habitation. It is a harmless shy nocturnal rodent that lives usually in dense low vegetation, and hides by day in a burrow. Its tunnel-like runways are often to be seen in low dense vegetation particularly around the borders of swamps and along creekbanks. This species is an inhabitant of coastal parts of Australia from north-eastern Queensland through eastern New South Wales to Victoria, South Australia and the south-west corner of Western Australia.

The Southern Bush-rat has fluffy-looking brown fur, and a tail about the same length or slightly shorter than the head-and-body. Incorporated within this species are the Allied Rat of coastal eastern Australia (formerly *Rattus assimilis*), Grey's Rat (formerly *R. greyi*) and the Western Swamp-rat *(R. fuscipes)*. Of these three sub-species, the Western Swamp-rat had been the first to be named. As it has now been discovered that there are insufficient differences between these three native rats to justify three separate species, all have been gathered together as a single species which must take the oldest of the names, that is, *fuscipes*; the other two become sub-species names, i.e. *Rattus fuscipes assimilis*, and *R. f. greyi*.

Spinifex Hopping-mouse   *Notomys alexis*

In general appearance these small mammals are like miniature kangaroos, but in fact are not marsupials, but rodents. They are about the size of an ordinary mouse or slightly larger, with very long kangaroo-like hind limbs, very short fore limbs, extremely long tufted tail, and long ears. There are about nine species, many so similar that exact identification is almost impossible except by scientific study of skulls and other anatomical features.

The hopping mice when moving fast travel in long kangaroo-like leaps. Their extremely long brush-tipped tails probably act as rudders to assist them to make sharp turns at speed — they are airborne for a greater time than touching the ground. When pursued they are fast and dodge effectively between the clumps of spiny spinifex.

This species inhabits the sandy spinifex grasslands and sand dune country of a great part of the arid interior of Australia, from western Queensland through central Australia to the interior and dry coastal parts of Western Australia. Spinifex Hopping-mice are able to survive in the semi-desert country because the hot days are spent in deep burrows a metre or more in depth, where the sand is always cool and often damp. Although they will drink water if it is available, their body chemistry is such that they can live and thrive on dry seed without any water or moist food at all.

Tunney's Rat   *Rattus tunneyi*

This attractive-looking native rat inhabits a wide variety of country including savannah-

woodlands, heaths, coastal sand flats and river plains. It occurs in northern Australia, from the mid west coast and northern parts of Western Australia through the Northern Territory to eastern Queensland and western New South Wales.

Also known as the Paler Field-rat, this species, about the size of an ordinary rat, has a very short tail, long soft fur giving a fluffy appearance. It is mostly sandy-buff in colour, except the underparts which are whitish or pale yellow. J. T. Tunney was an early collector or Australian fauna.

Water-rat   *Hydromys chrystogaster*

The Water-rat is the only Australian mammal apart from the Platypus to have become specialized for an aquatic way of life. These adaptations have occurred within Australia, making the species unique to this continent. It has partially webbed feet, short dense fur, flattened head with nostrils well up and forward, high-set eyes, and very small ears.

The Water-rat inhabits freshwater rivers and creeks almost throughout Australia, Tasmania and many offshore islands. Numerous

races occur, some of which in the past have been listed as district species — the Western Water-rat, Atherton Water-rat, and others. Some have attractively coloured fur. The Water-rat of eastern Australia has remarkable variability of the undersurface colour, from white to golden. The back may be dark grey, black, or golden brown.

Water-rats live in long burrows along river banks, with a nest chamber lined with twigs, grass and bark. In swamps they may live in old hollow logs, or in the massive floating nests of swans. Their food consists of mussels and other molluscs, and they will also kill young ducks and other water birds.

White-tipped Stick-nest Rat
Leporillus apicalis

The various stick-nest or house-building rats are gregarious, short-faced, fluffy-furred small rodents which construct large and solid structures of sticks, presumably as a defence against predators and for protection from the scorching desert sun. The architecture of the stick nests seems to depend largely upon the materials available and the local terrain. On the open plains the large sticks are interwoven around and among the lower branches of a low shrub, forming a compact structure about a metre in height, with a number of entrance holes around the base, and usually a shallow burrow beneath. Where there are no bushes to give a supporting framework the nests are a rather flattened pile of sticks. In rocky or rough country they may be built over an existing hole or crevice, and have stones incorporated among the sticks.

The White-tipped Stick-nest Rat is of similar size to an ordinary rat, with an untufted, white-tipped tail that is longer than the combined length of head and body. It inhabits the interior of Australia, from the desert heart of Western Australia to western New South Wales and north-western Victoria. It has become extremely rare, but may still exist in remote uninhabited desert regions.

South-west Pigmy Possum
Cercartetus concinnus

There are in Australia three species of pigmy possums (disregarding the very rare and unusual *Burramys*) which are all very much alike except for minor differences, such as size, or relative lengths of tails. In some instances they are more easily identified on the basis of locality. The South Western Pigmy Possum inhabits the forests, heathlands and scrublands of south-western New South Wales, western Victoria, southern and south-eastern parts of South Australia, and the south-west of Western Australia. Its distribution partly overlaps that of the Eastern Pigmy Possum (*C. nanus*) which occurs in the forests of coastal eastern and south eastern Australia and Tasmania. Also in Tasmania, but confined to that island and Kangaroo Island, is the slightly smaller Tasmanian Pigmy Possum (*C. lepidus*). Well separated geographically from the other species is the Long-tailed Pigmy Possum (*C. caudatus*) of north-eastern Queensland and New Guinea. Scientific study of dental features is the only really accurate means of identification.

These tiny mouse-sized possums are very similar to the preceding Feathertail Glider, without the gliding membranes and feathery-edged tail. The full prehensile tail is carried tightly rolled up like those of ringtail possums. The South-west Pigmy Possum is fawny or greyish brown, paler or white below. It is solitary, arboreal and nocturnal in habits, and lives upon insects and nectar. During the day it hides in disused birds nests, in small hollows of trees, skirts of blackboys or under loose slabs of bark. Pigmy Possums are among the few marsupials able to lower their body temperatures in cold weather and become torpid, a condition rather similar to hibernation.

PIGMY POSSUMS AND FEATHERTAIL GLIDER

FAMILY *BURRAMYIDAE*

The pigmy possum and pigmy glider are mouse-like in size and in their superficial appearance, but at closer inspection prove to be far more attractive little animals, with the big bulging black eyes typical of the nocturnal marsupials. They have pouches in which to carry their young, prehensile tails that can grip around twigs or leaves, and many small needle-like incisor teeth instead of the rodent mouse's typically chisel-like front teeth. These possums have the syndactyl toes (joined second and third toes on the hind feet with the double comb-like claws used in grooming the fur) which separates them not only from all ordinary mice, but also from the carnivorous marsupial mice which do not have this feature.

This family takes its names, Burramyidae, from a 'living fossil', the Mountain Possum (*Burramys parvus*). First discovered as fossil

bones in rock at the Wombeyan caves, New South Wales, in 1894, the *Burramys* was found to have teeth of a kind that seemed to relate it to the kangaroos. From these few bones it was not certain whether this animal was a kind of pygmy possum, or a miniature kangaroo. By a remarkable chance discovery the first living *Burramys* was found high in the Victorian Alps in 1966; until this time it was believed to have been extinct for many thousands of years. Now it is known that there are populations of these in various parts of the mountains of the south-east. The Mountain Pigmy Possum is somewhat larger than the other mouse-sized members of this family, which includes four species of pigmy possums (genus *Cercartetus*) and a single miniature glider, the Feathertail.

Feathertail Glider   *Acrobates pygmaeus*

This is the only mouse-sized glider; it resembles very closely the little pigmy possums, but is unique in having a feather-like tail. Along each side of the tail is a continuous row of long stiff hairs which make the tail wide and flat. The tail retains its prehensile ability, and the tiny possum

uses it to grasp leaves and twigs when climbing the foliage of trees. Along each side of its body, stretched tight between wrists and ankles when the glider holds outspread all four limbs, are the delicate gliding membranes. Like the larger gliders, the Feathertail jumps from the higher limbs and makes controlled gliding descents, dropping at an angle of about forty-five degrees and using the feather-like tail to steer, until it lands on the lower branches or trunk of the same or a nearby tree.

Feathertails inhabit the eucalypt forests and woodlands of eastern and south-eastern Australia from Cape York to Adelaide. At night these smallest of gliders hunt very actively for insects and raid the flowers of the gums and bottlebrushes for nectar. A globular nest of shredded bark and gum leaves is constructed in a hollow limb of a tree, and often houses a large family. Four young make the normal litter, and these are carried in the pouch on nocturnal rambles while still small.

PLATYPUS FAMILY

FAMILY *ORNITHORHYNCHIDAE*

The only species of the family is an animal highly

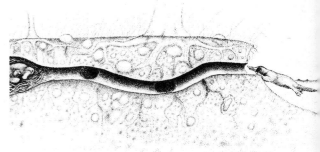

specialized for an aquatic way of life. It is one of the world's few living monotremes, primitive animals showing many of the reptile-like features carried on from the ancestors of present-day types of mammals. The female Platypus has no nipples, the milk being sucked up by the young as it seeps from the pore-like ducts of the mammary glands. Like the spiny anteaters, the platypus combines certain features of reptiles, marsupials and placental mammals. From the reptile skeleton they retain the pectoral or shoulder girdle bones and the reptile-like skull; they have a reptilian eye structure, a single combined genital and excretory opening (for which they are given the name 'monotreme') and they lay soft-shelled eggs like those of the reptiles. Characteristics shared by both monotremes and marsupials are the epipubic bones of the pelvic girdle, and the lack of any connections within the brain between the two cerebral hemispheres. Unlike reptiles, but in common with marsupial and placental mammals, they have mammary glands, a body covering of fur, a more efficient four-chambered heart, and a single lower jaw structure.

Features adapting the platypus to the aquatic environment are the webbed feet, flattened paddle-like tail, and superficially duck-like bill, extremely sensitive, for locating food underwater.

As there is but one species within the family, and as that species is confined to Australia the genus and family likewise are endemic.

Platypus *Ornithorhynchus anatinus*

This famed animal is often described as 'duck-billed', but this is a very superficial comparison. The beak or bill is not horny like those of birds, but is a frame of bone covered with soft moist skin, naked of any fur. In the water its webbed front feet expand into broad paddles. The hind feet, also webbed, trail behind. It has no external ears, and the eyes are small; both the ear and eye apertures are recessed into a deep groove in the skin, the edges of which close together underwater. This probably renders the Platypus both blind and deaf when beneath the surface, but it is in this environment that the marvellous bill takes over, guiding the Platypus along waterways and creekbeds, around boulders, under and over logs, among reeds, finding, identifying and capturing the prey. The diet consists of aquatic insect larvae, crustacea, worms, tadpoles and other small water creatures.

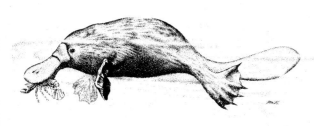

The Platypus inhabits freshwater streams and lakes of eastern Australia, from north-eastern Queensland through New South Wales and Victoria to south-eastern South Australia, and occurs also in Tasmania.

A burrow usually 4.5 to 9 metres, but sometimes up to 27 metres in length, is excavated by the female; the entrance is well above water level. At intervals along the tunnel are built plugs of soil, possibly to deter potential predators and to maintain a warm atmosphere for incubation of the eggs. In a chamber at the end a nest of

grass and gum leaves is constructed, and two soft-shelled eggs laid.

The monotremes have the distinction of being the only venomous mammals in Australia. The male Platypus has sharp venom spurs on both hind legs, and these inflict a painful wound, the effects of which last several days.

SEALS

FAMILY *OTARIIDAE*

Many marine mammals inhabit the oceans around the Australian coastline — seals, whales, porpoises and the dugong — but of these only the seals emerge from the sea to spend any considerable time on the shores of Australia or its offshore islands.

There are several groups of seals. The earless or true seals (family *Phocidae)* are the more completely adapted for aquatic life in having their hind limbs directed backwards instead of sideways; on land they can only manage a caterpillar-like wriggle. They wander the oceans and may not approach land for many weeks. As their name suggests there are no external ear-lobes. The Elephant Seal is one such earless seal of Australian waters.

Eared seals *(Otariidae)* are much more dependent upon land, and spend a great deal of time lying on rocks and beaches; the breeding places are on land, and the young spend their first weeks on the shores. Eared seals are of two types — the hair seals, and the fur seals. The former group includes the Australian Sea Lion, the latter the Australian Fur Seal and the New Zealand Fur Seal, all of which breed on Australian offshore islands, or are regular visitors to the coasts.

Australian Sea Lion *Neophoca cinerea*

The Sea Lion is one of the largest of all Australian mammals, the males exceeding three metres in length; it is surpassed in bulk only by the Elephant Seal, which is an occasional visitor to Tasmania and the south-eastern coastline.

The Sea Lion is one of the hair-seal group. Soon after birth the young lose their fine dense under-fur, retaining only the coarse hair coat. The males of this species can be recognized not only by their great size, but also by their pale yellowish manes, dog-like faces, very small ears,

and hind flippers which (unlike those of the Elephant Seal) can be turned forwards on land.

The Australian Sea Lion is a creature of Australia's southern oceans, from South Australia's Kangaroo Island right along the south coast into Western Australian waters, thence up the west coast as far as the Abrolhos Islands. It keeps mainly to the offshore islands. In spite of their awkward appearance and great size Sea Lions can move quite fast across a beach, are surprisingly agile in climbing over boulders, can even scale cliffs, and have been seen several miles inland. In the water they are swift and graceful.

SIMPLE-NOSED BATS

FAMILY *VESPERTILIONIDAE*

The small bats of this family are distinguished by the plain structure of the muzzle and lips; the complex nose-leaf structures of some other bats is absent, and the nostrils are situated at the extremity of the snout. The tail of these bats is characteristic, for it extends to the very edge of the tail membrane (which stretches between the two hind legs) but never more than a fraction of an inch beyond the edge of that membrane.

The bats of the world are divided into two major groups, the large mainly fruit-eating bats, flying foxes and blossom-bats (sub-order *Megachiroptera*), and the small insect-eaters (sub-order *Microchiroptera).* The latter group contains by far the majority of bats; they generally have wings which span less than one foot. The simple-nosed family of bats contains a great many of these. In Australia it includes the long-eared bats (genus *Nyctophilus*), the short-eared bats *(Eptesicus)*, the pipistrelles *(Pipistrellus),* the bent-wing bats *(Miniopterus),* the wattled bats *(Chalinolobus),* the simple bats *(Myotis),* the broad-nosed bats *(Nycticeius),* and dome-headed bats *(Phoniscus).*

Lesser Long-eared Bat
Nyctophilus geoffroyi

An inhabitant of the whole of southern and central Australia including the Northern Territory and Queensland north to the Gulf of Carpentaria, the Lesser Long-eared Bat shelters during the day in hollows of trees, or occasionally in rock crevices or caves. It may be found hiding under larger pieces of bark on a treetrunk, or even under a stone on the ground. It shares with the Greater Long-eared Bat, ears that join together over the forehead, and a nose leaf which is no more than several small rudimentary projections on the upper surface of the snout. The posterior part of the nose leaf is better developed into a Y shape than in the case of the Greater Long-eared Bat. This species is about nine centimetres in length. There are several other very similar species of this genus in tropical northern Australia.

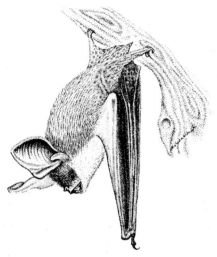

In early summer these bats form colonies containing both males and females. The two young immediately after birth cling to the undersurface of the mother and are carried by her on nightly insect-hunting flights. Very little is known about the behaviour of these bats at other times of the year; a limited amount of banding suggests that, in some districts at least, they are sedentary.

When hunting for insects, Lesser Long-eared Bats will land on the ground to take beetles and other insects. Their legs being too weak to assist takeoff they jerk themselves upwards by a flick of their wrists, wings folded, suddenly opening their wings when clear of the ground. The large ears are tightly folded while they sleep during the day.

SPINY ANTEATERS

FAMILY *TACHYGLOSSIDAE*

The spiny anteaters share with the Platypus the distinction of being a primitive link between the reptiles and the mammals. The spiny anteaters and Platypus are the only mammals to lay eggs, and have reptile-like features of reproductive system, skeleton, and body temperature regulation. One species occurs throughout Australia and New Guinea, and four species of another genus are found in New Guinea.

Within Australia the Spiny Anteater represents a most successful type of mammal in spite of its primitive structure. It has a wide distribution, and has remained common under conditions which have caused the loss of many of

the more advanced marsupial and placental mammals. Spiny Anteaters may be seen in a wide range of habitats, from semi-desert to rainforest.

The 'beak' which is common to all spiny anteater species is an adaptation to their way of feeding. The beak-like muzzle has at its tip a tiny mouth which opens just enough to allow the rapid probing action of the long tongue in picking up ants and termites.

Spiny Anteater *Tachyglossus aculeatus*

The one and only Australian species of its family, known as the Spiny Anteater or Echidna, is covered with long sharp spines, has a tubular beak-like snout, horny serrations on its tongue instead of teeth, and reproduces by laying eggs which are carried in a pouch that forms during the breeding season. It occurs throughout Australia, with a sub-species in Tasmania.

Confident of the protection given by its coat of spines, the anteater is out during the daytime, and is most active in early afternoon. If disturbed, its immediate reaction is to burrow straight downwards, and once half buried, with the heavy claws of its immensely strong limbs hooked under roots or rocks, it is almost impossible to dislodge. But if the anteater does happen to be taken by surprise on a hard surface where it cannot burrow or grip, it curls into a tight ball, protecting its face and underparts where there are no spines.

The female Spiny Anteater lays a single egg which is put in a temporary pouch. The anteater apparently lies on its back and curves its body so that the egg, perhaps with some guidance from the snout, rolls into the pouch. After seven to ten days in the pouch the egg hatches. The tiny hairless baby lives in the pouch for six to eight weeks, until its new spines become uncomfortable for its mother, who removes it to a safe hiding place until it is able to fend for itself.

WOMBATS

FAMILY *VOMBATIDAE*

Although still showing traces of tree-dwelling ancestors, the wombats have for so long been terrestrial animals that they have completely lost any useful tail they may have had. Their feeding habits have led to a remarkably rodent-like dentition — they differ from all other living marsupials in having only a single pair of upper and lower incisors. All the teeth grow continuously and are steadily worn away.

Wombats are extremely competent burrowers, constructing extensive and roomy net-

works of tunnels and subterranean chambers, often among and beneath boulders or tree roots. Their squat, solid build, powerful limbs and shovel-shaped nails enables them to excavate burrows up to a metre in diameter and as much as twenty metres in length. The ability of wombats to live without water derives partly from the cool damp environment of their burrows, where they can during daylight hours avoid all heat. However on cool sunny days they often sit or laze around near their burrow entrances.

Wombats, like other marsupials, have pouches, but these open towards the rear, thus being less likely to fill with sand. Australia has four species grouped in two genera: the forest wombats (genus *Vombatus*), and the plains wombats *(Lasiorhinus)*. The former genus, to which the Common and Tasmanian Wombats belong, inhabit forest and hilly country of the east and south-east. The plains (hairy-nosed) wombats are confined to smaller areas of open country, and are endangered by human activities.

Common Wombat *Vombatus ursinus*

The Common Wombat is a member of the forest or naked-nosed group. Also in this genus are the Tasmanian Wombat and the Islands Wombat (from the islands of Bass Strait). These can be recognized by their coarse thick body fur, short ears, and snout which has coarse granular skin without any fur. The Common Wombat is the mainland representative of the genus, and inhabits hilly or mountainous, usually rocky and forested country of coastal eastern Australia and the Great Dividing Range from southern Queensland and eastern New South Wales to Victoria and south-eastern parts of South Australia. It is the largest of the wombats, attaining a body length of well over one metre and of very heavy build.

Like other wombats, this species feeds upon grasses, tree roots and other plant material found in or below the ground.

The breeding season is about April to June, a single young is born, and carried in the pouch

until the following summer by which time it is fully furred. The young wombat is a typical marsupial in that it is born very immature and extremely small — only about two and a half centimetres in length. Immediately after birth it crawls through the fur to its mother's pouch, fastens to one of the two teats, and continues its development for several months before again showing itself to the outside world.

Hairy-nosed Wombat   *Lasiorhinus latifrons*

Unlike the Common Wombat this species has its snout covered with fine hair. Other differences are the much finer, more silken and glossy body fur, and the longer and more pointed ears. The several species of this group of wombats are known as plains wombats. Whereas the forests wombats keep to rocky hilly forested country, these inhabit drier inland plains. The Hairy-nosed Wombat's main area of distribution now is the Nullarbor Plains region, in the far south-western corner of South Australia and just across the border into Western Australia. Here it lives in a near desert environment on treeless limestone plains where there is rarely any surface water.

Hairy-nosed Wombats in times of drought are able to go without water for three or four months or more, but when rain does fall they drink from the small pools that collect on the plain's hard limestone capping. This drought-survival ability is due partly to their avoiding heat by living in deep burrows. These wombats have become very specialized in their feeding habits, preferring one type of perennial grass, but will eat other grasses at times of drought; unlike the common wombat they will not eat roots, bark or leaves.

The burrows of the Hairy-nosed Wombat are often clustered together to form extensive warrens, with entrances meeting together as a crater large enough for a small car to fall into.

Great Grey Kangaroo   *Macropus giganteus*
Australia's large kangaroos can be identified at a distance by the manner in which they travel. The Great Grey bounds along with the forequarters low but the head held up, and the tail curved upwards and pumped up and down in time with the bounding action of the powerful hind legs. The Red Kangaroo, however, travels with the head down, and the Wallaroo bounces along with the body rigidly upright. In places where they are not molested kangaroos come out of their shady resting places to feed in the late afternoon, and will be seen feeding until mid-morning in cool weather.

Spinifex Hopping Mouse *Notomys Alexis*
Hopping mice travel with Kangaroo like leaps. Their
long brush tipped tails probably act as rudders to assist
them to make sharp turns at speed. They are about the
size of an ordinary mouse.

Platypus *Ornithorhynchus anatinus*
Probably the strangest of all Australian animals, the
Platypus at first appears to be a peculiar mixture of
bird and mammal. But such unusual features as the
duck-bill and the webbed feet are adaptations to its
amphibious life, while the egg-laying way of
reproduction has been retained from its primitive
reptilian ancestors.

Koala *Phascolarctos cinereus*
Koalas are restricted to parts of eastern Australia
where suitable eucalypts grow, but their distribution,
even before their numbers were greatly reduced by
the activities of man, was much less than the range of
ácceptable tree species; fossils show that they once
occurred on the opposite side of the continent, in
Western Australia.

Tunney's Rat *Rattus tunneyi*
The sandy-brown, fluffy native rodent known as the
Paler Field-rat or Tunney's Rat is one of the more
attractive of Australian native rats. It keeps to its
native bushland haunts, and generally does not
interfere or conflict with man. Before the arrival of
European man, Australia contained more than fifty
species of native rats and mice.

Red Kangaroo *Megaleia rufa*
This inhabitant of the wide grassland plains of the vast
heartland of Australia, the big Red, or Plains Kangaroo,
has been driven back, exterminated from much of the
country it once inhabited. But elsewhere, partially as a
result of conservation legislation, it has remained
abundant. Preferring dry regions, it avoids the lush
tropical northern grasslands as well as the coastal
forests and the rocky ranges; those places have their
own kinds of kangaroos and wallabies.

Families in Zoological Order

Reptile families in conventional zoological order:

CHELONIIDAE	Turtles
DERMOCHELYIDAE	Leathery Turtle
CHELYIDAE	Tortoises
GEKKONIDAE	Geckoes
PYGOPODIDAE	Legless Lizards
AGAMIDAE	Dragons
SCINCIDAE	Skinks
VARANIDAE	Monitors or Goannas
TYPHLOPIDAE	Blind Snakes
BOIDAE	Pythons
COLUBRIDAE	Colubrid Snakes
ELAPIDAE	Elapid Snakes
HYDROPHIIDAE	Sea Snakes
CROCODYLIDAE	Crocodiles

Bird families in conventional zoological order:

CASUARIIDAE	Cassowaries
DROMAIIDAE	Emus
SPHENISCIDAE	Penguins
DIOMEDEIDAE	Albatrosses
PROCELLARIIDAE	Petrels, Shearwaters, Prions
HYDROBATIDAE	Storm-petrels
PELECANOIDIDAE	Diving Petrels
PODICIPEDIDAE	Grebes
PELECANIDAE	Pelicans
SULIDAE	Gannets and Boobies
PHALACROCORACIDAE	Cormorants and Darter
FREGATIDAE	Frigate birds
PHAETHONTIDAE	Tropic-birds
ARDEIDAE	Bitterns, Herons and Egrets
CICONIIDAE	Storks
THRESKIORNITHIDAE	Ibis and Spoonbills
ANATIDAE	Swans, Geese and Ducks
ACCIPITRIDAE	Hawks and Eagles
PANDIONIDAE	The Osprey
FALCONIDAE	Falcons
MEGAPODIIDAE	Mound Builders
PHASIANIDAE	Pheasants and Quails
TURNICIDAE	Button Quails and Plains Wanderer
GRUIDAE	Cranes
RALLIDAE	Rails and Crakes
OTIDAE	Bustards
JACANIDAE	Jacanas
ROSTRATULIDAE	The Painted Snipe

Australian Sea Lion *Neophoca cinerea*
On islands along Australia's southern coastline Sea Lions come ashore to breed between October and December, each female giving birth to a single pup. These seals congregate above the beaches where each of the massive bulls establishes a harem. The pup hides in crevices and pools among the rocks while the female is absent, and she calls it from hiding when she returns from the sea.

HAEMATOPODIDAE	Oystercatchers
CHARADRIIDAE	Plovers and Dotterels
SCOLOPACIDAE	Sandpipers, Snipe, Curlews, etc.
PHALAROPODIDAE	Phalaropes
RECURVIROSTRIDAE	Avocets and Stilts
BURHINIDAE	Stone Curlews
GLAREOLIDAE	Pratincoles and Coursers
STERCORARIIDAE	Skuas
LARIDAE	Gulls, Noddies and Terns
COLUMBIDAE	Pigeons and Doves
PSITTACIDAE	Lorikeets, Cockatoos and Parrots
CUCULIDAE	Cuckoos and Coucals
TYTONIDAE	Barn Owls
STRIGIDAE	Typical Owls
PODARGIDAE	Frogmouths
AEGOTHELIDAE	Owlet Nightjars
CAPRIMULGIDAE	Nightjars
APODIDAE	Swifts
ALCEDINIDAE	Kingfishers
MEROPIDAE	Bee-eaters
CORACIIDAE	Rollers
PITTIDAE	Pittas
MENURIDAE	Lyrebirds
ATRICHORNITHIDAE	Scrub-birds
ALAUDIDAE	Larks
HIRUNDINIDAE	Swallows and Martins
MOTACILLIDAE	Pipits and Wagtails
CAMPEPHAGIDAE	Cuckoo-shrikes and Trillers
TURDIDAE	The Thrushes
TIMALIIDAE	Babblers
MALURIDAE	Fairy Wrens, Emu-wrens, and Grass-wrens
SYLVIIDAE	Warblers
MUSCICAPIDAE	Flycatchers and Robins
MONARCHIDAE	Monarch and Allied Flycatchers
PACHYCEPHALIDAE	Whistlers and Thrushes
EPHTHIANURIDAE	The Chats
NEOSITTIDAE	The Sittellas
CLIMACTERIDAE	Treecreepers
DICAEIDAE	Flowerpeckers
NECTARINIIDAE	Sunbirds
ZOSTEROPIDAE	Silvereyes
MELIPHAGIDAE	Honeyeaters
ESTRILDIDAE	Finches
PLOCEIDAE	Weaver-birds
FRINGILLIDAE	Finches
STURNIDAE	Starlings
ORIOLIDAE	Orioles
DICRURIDAE	The Drongos
GRALLINIDAE	The Mud-nest Builders
ARTAMIDAE	Woodswallows
CRACTICIDAE	Butcher-birds Magpies, Currawongs
PTILONORHYNCHIDAE	Bowerbirds
PARADISAEIDAE	Birds of Paradise
CORVIDAE	Crows and Ravens

Mammal families in conventional zoological order:

MACROPODIDAE	Kangaroos
PHALANGERIDAE	Cuscuses and Large Possums
PETAURIDAE	Ringtail Possums and Large Gliders
BURRAMYIDAE	Small Possums and Gliders
TARSIPEDIDAE	Honey Possum
PHASCOLARCTIDAE	Koala
VOMBATIDAE	Wombats
PERAMELIDAE	Bandicoots
DASYURIDAE	Native Cats and other marsupial carnivores
THYLACININAE	Tasmanian Tiger
NOTORYCTIDAE	Marsupial-mole
MURIDAE	Native Rats and Mice
MEGADERMATIDAE	Ghost Bat
VESPERTILIONIDAE	Simple-nosed Bats
RHINOLOPHIDAE	Horseshoe Bats
HIPPOSIDERIDAE	Horseshoe Bats
MOLOSSIDAE	Mastiff Bats
EMBALLONURIDAE	Sheath-tailed or Free-tailed Bats
PTEROPODIDAE	Flying Foxes, Fruit and Blossom Bats
CANIDAE	Dingoes
OTARIIDAE	Seals
PHOCIDAE	Earless or True Seals
TACHYGLOSSIDAE	Spiny Anteater
ORNITHORHYNCHIDAE	Platypus

Index of Scientific Names

Index of Common Names

Index of Family Names

Further Reading

What Bird is That?	Caley, N. W.	Angus and Robertson, Sydney, 1973
Australian Reptiles in Colour	Cogger, H.	A. H. & A. W. Reed, Sydney, 1967
Nightwatchmen of Bush and Plain	Fleay, D.	Jacaranda Press, Brisbane, 1968
Kangaroos	Frith, H. J. & Calaby, J. H.	F. W. Cheshire, Melbourne, 1968
The Mallee Fowl	Frith, H. J.	Angus and Robertson, Sydney, 1962
Waterfowl in Australia	Frith, H. J.	Angus and Robertson, Sydney, 1967
Australian Parrots	Forshaw, J. M.	Lansdowne Press, Melbourne, 1969
A Handbook of the Snakes of Western Australia	Glauert, L.	W. A. Naturalists Club, Perth, 1957
The Snakes of Australia	Kinghorn, J. R.	Angus and Robertson, Sydney, 1957
Birds of Australia	Macdonald, J. D.	A. H. & A. W. Reed, Sydney, 1973
Bower Birds	Marshall, A. J.	Oxford University Press, New York, 1954
Australian Marsupial and Other Native Mammals	Morcombe, M. K.	Lansdowne Press, Melbourne, 1972
Birds of Australia	Morcombe, M. K.	Lansdowne Press, Melbourne, 1971
Bush Birds in Colour	Morcombe, M. K.	A. H. & A. W. Reed, Sydney, 1974
Australian Flycatchers and their Allies	Officer, H. R.	The Bird Observer's Club, Melbourne, 1969
Australian Honeyeaters	Officer, H. R.	The Bird Observer's Club, Melbourne, 1964
A Guide to the Native Mammals of Australia	Ride, W. D. L.	Oxford University Press, Melbourne, 1970
Handbook of Australian Seabirds	Serventy, D. L. & Others	A. H. & A. W. Reed, Sydney, 1971
Birds of Western Australia	Serventy, D. L. & Whittel, H. M.	Paterson Brokensha, Perth, 1962
A Field Guide to Australian Birds	Slater, P.	Rigby, Adelaide, 1970
The Lyrebird	Smith, L. H.	Lansdowne Press, Melbourne, 1970
Furred Animals of Australia	Troughton, E.	Angus and Robertson, Sydney, 1948
Dangerous Snakes of Australia and New Guinea	Worrel, E.	Angus and Robertson, Sydney, 1965
Reptiles of Australia	Worrel, E.	Angus and Robertson, Sydney, 1963